Engineering
Women and Leadership

© Springer Nature Switzerland AG 2022

Reprint of original edition © Morgan & Claypool 2008

All rights reserved. No part of this publication may be reproduced, stored in a retrieval system, or transmitted in any form or by any means—electronic, mechanical, photocopy, recording, or any other except for brief quotations in printed reviews, without the prior permission of the publisher.

Engineering: Women and Leadership
Corri Zoli, Shobha Bhatia, Valerie Davidson, and Kelly Rusch

ISBN: 978-3-031-79945-7   paperback

ISBN: 978-3-031-79946-4   ebook

DOI: 10.1007/978-3-031-79946-4

A Publication in the Springer series

SYNTHESIS LECTURES ON ENGINEERING, TECHNOLOGY AND SOCIETY #5

Lecture #5

Series Editor: Caroline Baillie, Queens University

Series ISSN

ISSN 1933-3633   print

ISSN 1933-3461   electronic

Engineering
Women and Leadership

Corri Zoli
Syracuse University

Shobha Bhatia
Syracuse University

Valerie Davidson
University of Guelph and National Science and Engineering Research Council of Canada

Kelly Rusch
Louisiana State University

SYNTHESIS LECTURES ON ENGINEERING, TECHNOLOGY AND SOCIETY #5

ABSTRACT

In this book we explore a sea change occurring in leadership for academic women in the sciences and engineering. Our approach is a two-pronged one: On the one hand, we outline the nature of the changes and their sources, both in various literatures and from program research results. On the other hand, we specify and provide detail about the persistent problems and obstacles that remain as barriers to women's full participation in academic science and engineering, their career advancement and success, and, most important, their role as leaders in making change. At the heart of this book is our goal to give some shape to the research, practice, and programs developed by women academic leaders making institutional change in the sciences and engineering.

KEYWORDS

Engineering, women, leadership, academic leadership, diversity, education, institutional transformation, and change

Acknowledgments

The authors would like to gratefully acknowledge support received from the National Science Foundation and the Women in Engineering Leadership Institute, as well as Syracuse University's L.C. Smith College of Engineering and Computer Science and the Graduate School. The authors would also like to thank Debbie Niemeier at the University of California-Davis for her role in assessment and evaluation and Tracey Manning at the University of Maryland-College Park for her role in leadership coaching and training. The authors would also like to express their sincere thanks to all participating members of the WELI community, including conference attendees and speakers.

Contents

1. **Women in a New Era of Academic Leadership** 1
   1.1 Introduction... 1
   1.2 Broader Cultural Shifts 2
   1.3 Few Advanced Women in Science and Engineering 3
   1.4 Summary: Leadership in Making Change 4
   1.5 Prioritizing Questions 7

2. **Background: Academic Leadership for Women in Science and Engineering** 9
   2.1 Introduction... 9
   2.2 Accomplishments 9
   2.3 Remaining Challenges.. 14
       2.3.1 Gendered Field Segregation and Advancement 14
       2.3.2 Disaggregating "Diversity".. 17
       2.3.3 Transforming S&E Climates.. 20
       2.3.4 Fully Utilizing the Pool of Qualified, Talented S&E Candidates.......... 22
       2.3.5 Leadership: Promoting Women's Authority in S&E Education 22

3. **Gender and Leadership: Theories and Applications** 27
   3.1 Introduction... 27
   3.2 Leadership Typologies 27
   3.3 Leading Change 30
   3.4 Women and Leadership 32
       3.4.1 The Female Advantage? Gender, Style, and Effectiveness.................... 34
       3.4.2 Prejudice, Performance, and Assumptions............................ 37
   3.5 Strategic Engagements in S&E... 40
   3.6 Women as Change Agents 43

4. **Women in Engineering Leadership Institute: Critical Issues
   for Women Academic Engineers as Leaders** 45
   4.1 Introduction: Identifying Critical Issues..................................... 45
   4.2 Definitions: Multifaceted Leadership in Academe............................ 46

4.3   Institutional Transformation: NSF-ADVANCE and Other
Leadership Programs... 47

4.4   The Women in Engineering Leadership Institute.................................. 52

4.4.1   WELI's Unique Format... 55

4.5   Building an Alternative Network 60

4.6   Transformative Theories of Leadership... 64

5.   **From Success Stories to Success Strategies: Leadership for Promoting Diversity
in Academic Science and Engineering** 69

5.1   Introduction: Success Strategies and "Insider Knowledge"...................... 69

5.2   WELI Advanced Leadership & Institutional Transformation Conference....... 70

5.2.1   Twin Goals: Individual Advancement and Institutional Change 70

5.2.2   Action-oriented Roadmap... 72

5.2.3   Mapping Out Individual Goals in Context............................ 72

5.2.4   Strategies: Individual to Institutional Success 73

5.3   Assessment Results: WELI Advanced Leadership & Institutional
Transformation Conference... 73

5.4   Success-Oriented Principles for Diversity Leadership 76

5.4.1   Principles for Diversity Leadership in the
Academic S&E Context... 77

5.4.2   Future Research 79

6.   **Conclusion** 81

References 83

Author Biography 105

CHAPTER 1

Women in a New Era of Academic Leadership

1.1   INTRODUCTION

Something exciting is happening for women in academia, particularly in science and engineering (S&E). Harvard, MIT, Princeton, Brown, University of Pennsylvania, Ohio State University, University of Michigan, RPI, Case Western Reserve, Syracuse University, Lehigh University, CUNY-Hunter College among others, all boast female presidents—many of whom are scientists and engineers.[1] Not only is Harvard's newest president Drew Gilpin Faust this institution's first woman president, but she emerged while dean of the Radcliffe Institute for Advanced Study on Women, Gender, and Society in charge of Harvard's twin Task Forces on Women Faculty and Women in Science and Engineering (2005)—each formed to develop concrete solutions for reducing barriers to women's advancement at Harvard and in women's academic careers more broadly.[2]

In certain respects, this admittedly small but powerful group of prominent, visionary leaders represents a trend in academic leadership at least ten years in the making. In addition to a groundswell of new and important interdisciplinary scholarship on women, gender, and leadership over the last decade, there is a concerted effort, especially at academic institutions, to notice whether women are occupying leadership positions (i.e., president, provost, dean, department chair, research

---

[1] Historian Drew Gilpin Faust, Harvard University; neurobiologist Susan Hockfield, Massachusetts Institute of Technology; molecular biologist Shirley Tilghman, Princeton University; French literary scholar Ruth Simmons, Brown University; political scientist Amy Gutmann, University of Pennsylvania; biologist Karen Holbrook, Ohio State University; biochemist Mary Sue Coleman, University of Michigan; theoretical physicist Shirley Jackson, Rensselaer Polytechnic Institute; professor of law Barbara Snyder, Case Western Reserve; Social Psychologist Nancy Cantor, Syracuse University; chemical engineer Alice Gast, Lehigh University; public affairs professional Jennifer J. Raab, Hunter College of The City University of New York.

[2] For Lawrence Summers's original remarks about women's aptitude in science and engineering and the responses, see Women in Science and Engineering Leadership Institute (WISELI), Resources on the Lawrence Summers Debate, University of Wisconsin-Madison, http://wiseli.engr.wisc.edu/news/Summers.htm#LS.

director, and non-titled positions) and how they are faring. Add to these efforts the National Science Foundation's (NSF) leadership-focused Project ADVANCE to "increase the participation and advancement of women in academic science and engineering careers," and we are witnessing a sea change in academic leadership—in women's leadership development and training, in women's contribution to new models of leadership style, and in women's advancement and promotion.[3] Related efforts, the NSF-sponsored Women in Engineering Leadership Institute (WELI) and the Committee on the Advancement of Women Chemists (COACh), have even begun to adapt leadership development to field-specific needs and challenges.

1.2 BROADER CULTURAL SHIFTS

These changes echo broader cultural shifts that recognize women's potential as leaders, the importance of resourcing women to achieve leadership positions, and a growing sense that industries, sectors, even nations cannot survive without utilizing half the available talent pool. Well-known global investment firm Goldman Sachs recently announced a $100 million donation to develop the business and leadership capacity of 10,000 women in Africa, the Middle East, across the developing world, and in the United States, for instance, with these cautionary words: "No country will ever achieve its full potential if half of its talent pool is stymied or underrepresented."[4]

But it is important to temper this excitement with a sober examination of these changes—how deep do they go? Are they sustainable? Who is included and excluded? How serious are the institutional commitments? In certain respects, the high-profile appointments of women presidents at especially Ivy League and research institutions do represent some change, on the one hand, but a thin veneer of professional success and institutional transformation, on the other. As the American Council of Education's 2007 edition of *The American College President* notes, while the percentage of women university and college presidents has more than doubled from 9.5 to 23% from 1986 to 2006, growth has slowed in recent years, as percentages of women presidents remained at 19.3% in 1998 and 21.1% in 2001.[5] Moreover, women are least likely to head doctorate-granting institutions: they made up 13.8% at these institutions in 2006, up only slightly from 13.3% in 2001, and 13.2% in 1998. Equally important, the percentage of presidents who are members of ethnic/racial minority groups has likewise grown from 8.1% to 13.6% during this same period of 1986–2006; yet, this

---

[3] See NSF 07-582 "Increasing the Participation and Advancement of Women in Academic Science and Engineering Careers (ADVANCE)," www.nsf.gov/pubs/2007/nsf07582/nsf07582.pdf, 2.

[4] Stephanie Strom, "Gift to Teach Business to Third-World Women," *New York Times*, March 6, 2008.

[5] Jacqueline King and Gigi Gomez, *The American College President: 2007 Edition*, Report 6, The American College President Study (ACPS) Series (Washington DC: American Council on Education, 2007).

group's growth has also slowed, with minority presidents ranging from 10.7% in 1995, 11.3% in 1998, and 12.8% in 2001.

1.3 FEW ADVANCED WOMEN IN SCIENCE AND ENGINEERING

Many studies, it is worth noting, have observed a dearth of role models and mentors for women; or looked at how women are often subtly excluded from career advancement networks and informal informational sharing sessions; or even observed the infrequency with which women are supported through intermediary leadership positions (department chair, for instance) as part of the valuable process of gaining leadership experience.[6] On this last point, women are often rushed into leadership positions without adequate support or preparation, which can then result in performance challenges that appear to confirm gender stereotypes about women leaders. These and additional issues represent a now well-established body of research that has uncovered persistent barriers—albeit often complex and subtle ones—to women's advancement, especially in predominantly male-dominated fields.[7] The glaring discrepancy between the percentage of women earning Ph.D.'s and those achieving faculty positions at top universities in S&E, for instance, provides a sobering indication of the pervasiveness of institutional obstacles in the academic sector (see Table 1.1 and Figure 1.1).

The severity of this discrepancy has prompted many of the university presidents already mentioned to publicly advocate for making good on the goal of "equal opportunity in education, training, and employment in scientific and technical fields" required by U.S. statute and to more "fully utiliz[e] the pool of women scientists they have produced."[8]

[6] N.C. Chesler and M. A. Chesler, "Gender-Informed Mentoring Strategies for Women Engineering Scholars: On Establishing a Caring Community," *Journal of Engineering Education* 91.1 (Jan. 2002): 49–55; J.M. Cain et al., "Effects of Perceptions and Mentorship on Pursuing a Career in Academic Medicine in Obstetrics and Gynecology," *Academic Medicine* 76.6 (June 2001): 628–634; L.L. Bakken, "Who Are Physician-Scientists' Role Models? Gender Makes a Difference," *Academic Medicine* 80.5 (May 2005): 502–506; P.L. Carr, et al., "A Ton of Feathers: Gender Discrimination in Academic Medical Careers and How to Manage It," *Journal of Women's Health* 12.10 (Nov. 2003): 1009–1018; C.Q. Choi, "Women Scientists Face Problems," *The Scientist* 18.3 (Feb. 2004); D.A. Niemeier and C. González, "Breaking into the Guildmasters' Club: What We Know About Women Science and Engineering Department Chairs at AAU Universities," *NWSA Journal* 16.1 (2004): 157–171.

[7] Committee on Maximizing the Potential of Women in Academic Science and Engineering, National Academy of Science, National Academy of Engineering, Institute of Medicine, *Beyond Bias and Barriers: Fulfilling the Potential of Women in Academic Science and Engineering* (National Academies Press: Washington, D.C., 2007); S.V. Rosser and E. O'Neil Lane, "Key Barriers for Academic Institutions Seeking to Retain Female Scientists and Engineers: Family-Unfriendly Policies, Low Numbers, Stereotypes, and Harassment," *Journal of Women and Minorities in Science and Engineering* 8 (2002): 161–189.

[8] J. Handelsman, N. Cantor, M. Carnes, D. Denton, E. Fine, B. Grosz, V. Hinshaw, C. Marrett, S. Rosser, D. Shalala, and J. Sheridan, 2005, "More Women in Science," *Science* 309: 1190–1199, www.sciencemag.org/cgi/content/full/309/5738/1190.

TABLE 1.1: Women Ph.D.'s and faculty for the top 50 departments in selected S&E disciplines

DISCIPLINE	CAREER LEVEL (PERCENT WOMEN)			
	PH.D.	ASST. PROF.	ASSOC. PROF.	FULL PROF.
Biology	45.89	30.20	24.87	14.79
Physical science (overall)	24.68	16.13	14.18	6.36
Astronomy	22.88	20.18	15.69	9.75
Chemistry	33.42	21.47	20.50	7.62
Computer science	15.27	10.82	14.41	8.33
Math and statistics	26.90	19.60	13.19	4.56
Physics	14.78	11.15	9.41	5.24
Engineering (overall)	15.34	16.94	11.17	3.68
Electrical	12.13	10.86	9.84	3.85
Civil	17.90	22.26	11.50	3.52
Mechanical	10.93	15.65	8.89	3.17
Chemical	24.98	21.38	19.19	4.37

Data on Ph.D.'s and faculty come from the same "top 50" departments for each discipline (ranked by NSF according to research expenditures in that discipline) and detailed by Nelson (2005). Ph.D. data are from 2001 to 2003 (NSF 2007); faculty data are from 2002, except Astronomy 2004 and Chemistry 2003 (Nelson 2005).

Source: J. Handelsman et al. (2005) [158].

1.4   SUMMARY: LEADERSHIP IN MAKING CHANGE

In this book we explore one dimension of this goal: the role of leadership for making change. It is now commonplace that leadership in general, not only in higher education, is a prerequisite to effect broad-based, sustainable institutional transformation. But while much has been written about women in S&E in professional, educational, historical, and industry contexts and with a focus on K-12, undergraduate, and graduate students, only recently has attention been paid to women S&E faculty as leaders, including their steps to advancement.

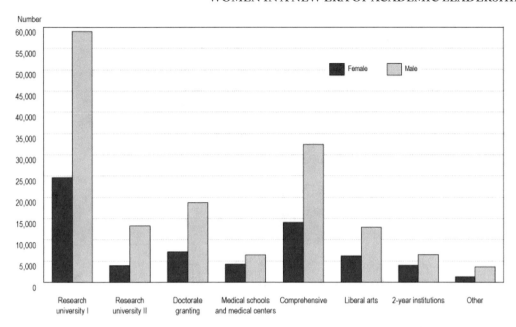

FIGURE 1.1: Doctoral S&E faculty by gender for Carnegie Classification, 2003. This includes full, associate, and assistant professors and instructors. Source: National Science Foundation, Division of Science Resources Statistics, Scientists and Engineers Statistical Data System (SESTAT).

According to the National Academies 2007 Report *Beyond Bias and Barriers: Fulfilling the Potential of Women in Academic Science and Engineering*, for instance, not only do the small proportion of women S&E faculty members at research universities "typically receive fewer resources and less support than their male colleagues," the "representation of women in leadership positions in our academic institutions, scientific and professional societies, and honorary organizations is low relative to the numbers of women qualified to hold these positions."[9] The *Report* (2007, 2) further notes that "it is not lack of talent, but unintentional biases and outmoded institutional structures" that hinder women—an "underuse of precious human capital" which "neither our academic institutions nor our nation can afford." Likewise, the recent BEST Report (2004), *The Talent Imperative: Diversifying America's Science and Engineering Workforce*, argues that institutional leadership "matters in creating successful programs" to "broaden the participation of women, underrepresented minorities, and persons with disabilities in science, engineering, and technology," and that this "commitment by

---

[9] COSEPUP, NAS, MAE, IOM, *Beyond Bias and Barriers*, 2.

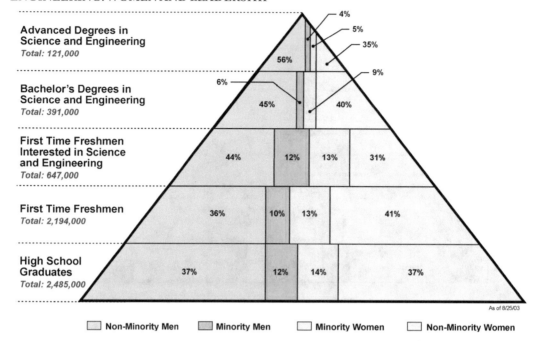

FIGURE 1.2: Talent lost in S&E from high school to advanced degrees. "Minority" means African American, Latino, and Native American. Source: Joan Burrelli, NSF, based on 1999 Common Core of Data, U.S. Department of Education, National Center for Education Statistics (NCES); NCES, 1998 IPEDS Fall Enrollment Survey; UCLA Higher Education Research Institute, 1998 American Freshman Survey (estimate); and NCES, 1998 IPEDS Completions Survey (Best 2004b, 20).

administration and senior faculty helps to ensure that increasing participation is an essential part of successful higher education programs."[10]

Taking into consideration these premises that institutional barriers constrict the use of our available talent pool and that successful leadership programs help to rectify this problem, our approach is a two-pronged one. On the one hand, we outline the nature of the changes and their sources in a now vibrant interdisciplinary literature on gender, leadership, and organizational change and from the research results of leadership-focused women in S&E programs at academic institutions. This research has importantly provided resources for women leaders in facilitating academic and institutional change. On the other hand, we specify and provide detail about the persistent problems and obstacles that remain as barriers to women's full participation in academic S&E,

[10] The Best Initiative, *The Talent Imperative: Diversifying America's Science and Engineering Workforce*, www.bestworkforce.org/PDFdocs/BESTTalentImperative_FullReport.pdf (San Diego, CA, 2004b): 9, 20.

including the newest challenges faced by women en route to advancement in the technical fields. At the heart of this book is our aim to give some shape to the research, practice, and programs developed by women academic leaders as activists in S&E so that what is still a veneer of professional success and leadership can become more integrative—with women's participation in S&E occurring at all levels, including in the pipeline leading to professional advancement.

Toward these ends, we divide this book into five chapters: the remainder of Chapter 1 provides an introduction to our approach. Chapter 2 presents a background on the problem of women's low numbers in academic S&E, the gains made, and how the newest research into this problem defines the challenges of the future. Chapter 3 addresses interdisciplinary literatures on women, gender, and leadership. Chapter 4 investigates the data from NSF-ADVANCE grants and other academic leadership programs, including the Women in Engineering Leadership Institute (WELI), some of the most critical issues for women scientists and engineers as academic leaders, and, from this context, key themes for women across available best practices, training modules, leadership training curriculums, and pedagogies. Chapter 5 presents our conclusion, which focuses on the limitations of these data, new and pressing research questions, and where advocates of this critical need area can go next.

1.5   PRIORITIZING QUESTIONS

It is worth briefly relaying our approach to this area of research and institutional practice. In the history and philosophy of science, Thomas Kuhn has argued that scientific advances occur, not by a linear accumulation of new knowledge, as one might assume, but by episodic "revolutions," what he termed "paradigm shifts," changes in our most basic assumptions under a dominant theory, including the rules and procedures specific to a field of inquiry—all of which may take the form of questions.[11] One aspect of the paradigm shift relevant to the study of women in the science, technology, engineering, and mathematics (or STEM) fields is the generative role of questions, especially new questions, in orienting change. Throughout this book, we highlight the questions galvanizing and organizing the changing academic culture of higher education, the new and critical questions asked by researchers in studying these developments, and the role of women scientists and engineers in raising questions by processing their own experiences in academia and in academic leadership.

Some paradigmatic questions, for instance, include:

1.   *Matters of relevance*: How are women academic leaders in S&E not simply a "special interest" problem, relevant to a small portion of the university community, but indicative of a broader cultural change in academic institutions and in contemporary society, i.e., changes

---

[11] Thomas Kuhn, *The Structure of Scientific Revolutions* (Chicago: University of Chicago Press, 1962) 109.

in leadership norms, styles and roles across sectors, demographic changes, transformations in gender relations, new standards of public accountability in academic institutions, and changes in the social role of engineering?

2. *Pragmatic matters*: Why is leadership an important mechanism to alleviate the underused talent of women and minorities in the technical fields? This question involves "critical mass" arguments, the premise that leadership may function as a tool for institutional transformation, and the role of gender in considering new and potentially more effective models of leadership supportive of new institutions.

3. *Theoretical matters*: Why is the field of women and leadership necessarily interdisciplinary and what are the challenges of conducting interdisciplinary inquiry in the technical fields? How are academic excellence and arguments for the educational benefits of diversity increasingly important in thinking about this particular constituency?

4. *Bridging the gap*: Most self-reflexive methodologies in recent years refute the separation between "theory" and "practice," whether in the human and social sciences or in the natural sciences. In this respect, we also try to "bridge this gap" by constantly integrating and synthesizing the newest research on women in leadership with program results from women in S&E programs, including our programmatic experiences across our collaborative group.

Each of these question clusters attempts to offer readers a frame of reference for approaching the following discussion of a highly dynamic area in higher education and the emergence of a critical consciousness among women scientists and engineers in their activist roles in leading change.

CHAPTER 2

Background: Academic Leadership for Women in Science and Engineering

2.1 INTRODUCTION

Understanding academic leadership, including how the "lessons learned" from women in the academic sciences and engineering offers broader insights into leadership more generally, requires surveying the accomplishments and challenges for this particularly group. In this chapter we first provide an overview of the increase in numbers of women in S&E at virtually every level of higher education over the last several decades: among undergraduate and graduate students, including masters and doctorate recipients, and among faculty at various levels. We then show how, despite this increase, numbers of women at the faculty level and among academic leaders—the focus of this book—remain small, especially at research institutions. This dearth of senior women has a ripple effect: it means that isolation and few mentors will continue at all levels and that the pool of available women to undertake academic leadership positions will be equally small. We argue, however, that while an emphasis on numbers, on increasing representation, is critical, it does not tell the whole story about the challenges faced by women faculty and academic leaders in S&E. We, thus, identify five remaining challenges for women in these academic fields by synthesizing recent research with experiences, including our own, from women in S&E programs.

2.2 ACCOMPLISHMENTS

Today's impressive incursion of women and minorities into S&E is a result of the rapid social changes of the late 1960s and 1970s. As Figures 2.1 and 2.2 show, S&E bachelor's and master's degrees awarded to women have increased every year since 1966 (with the exception of 1988), so that in 2005 women earned 45% of S&E (and 61% of non-S&E) doctoral degrees—up from 8% and 18%, respectively, in 1966. Likewise, S&E bachelor's and master's degrees also increased for minority groups since 1983 and 1990—with the proportion of master's degrees earned more than doubling between 1990 and 2004 from 6% to 16%. In the case of earned doctorates, as Figure 2.3 indicates, women's numbers (U.S. citizen and permanent residents) also rose from 785 in

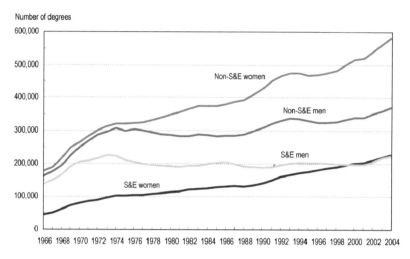

FIGURE 2.1: Bachelor's degrees awarded in S&E and non-S&E fields by sex: 1966–2004. National data are not available for 1999. Source: National Science Foundation, Division of Science Resources Statistics, special tabulations of U.S. Department of Education, National Center for Education Statistics, Integrated Postsecondary Education Data System, Completions Survey, 1966–2004, http://www.nsf.gov/statistics/wmpd/figc-1.htm.

1966 to 6,913 in 1995 and remains at approximately 7,000 doctorates per year through 2005. It is important to note that the number of earned doctorates by men (U.S. citizen and permanent resident) has simultaneously dropped from 1971 through the late 1980s and again from 1995 through 2002.[1]

These changes may be attributed to the intense cultural transformations and reform-oriented social movements associated with women's liberation, civil rights, and various student and education movements of the same period. But legal provisions, a sign of the times, also played no small role. Title VII of the Civil Rights Act of 1964 (Pub. L. 88-352), as amended by the Equal Employment Opportunity Act of 1972 (Pub. L. 92-261), which strengthened Title VII; Title IX of the Education Amendments of 1972 (U.S.C. Sections 1681–1688); and the 1972 Higher Education Affirmative Action Guidelines, for instance, all set out important standards against discrimination based on gender, race, creed, and national origin in employment and education among federally funded

---

[1] National Science Foundation, Division of Science Resources Statistics, *Women, Minorities, and Persons with Disabilities in Science and Engineering: 2004*, NSF 04-317 (Arlington, VA, 2004), www.nsf.gov/statistics/women.

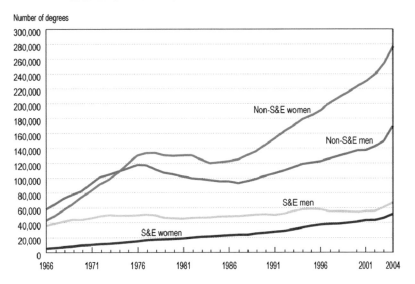

FIGURE 2.2: Master's degrees awarded in S&E and non-S&E fields by sex: 1966–2004. National data are not available for 1999. Source: National Science Foundation, Division of Science Resources Statistics, special tabulations of U.S. Department of Education, National Center for Education Statistics, Integrated Postsecondary Education Data System, Completions Survey, 1966–2004, http://www.nsf .gov/statistics/wmpd/figd-1.htm.

programs.[2] In many respects, socio-cultural and legal developments were intimately linked, as this "revolution in the laws governing women's education and employment rights" fueled and was fueled by "dramatic changes in our society's view of the role of women at home and in the workplace."[3] Most notably in 1980, the U.S. Congress passed the *Science and Engineering Equal Opportunities Act* (Public L. 96-516), which formalized into a "policy of the United States" that "men and women have equal opportunity in education, training, and employment in scientific and technical fields" and further set out concrete "guidelines for activities carried out pursuant to this Act."[4]

[2] For a full description of Pub. L. 88-352, see www.eeoc.gov/policy/vii.html; Title 20 U.S.C. Sections 1681–1688, see www.dol.gov/oasam/regs/statutes/titleix.htm; and Pub. L. 92-261, see www.eeoc.gov/abouteeoc/35th/thelaw/ eeo_1972.html.

[3] J. Scott Long, ed., *From Scarcity to Visibility: Gender Differences in the Careers of Doctoral Scientists and Engineers*, (Washington, D.C.: National Academy Press, 2001): 9; Rossiter addresses, however, the limits of this otherwise "legal revolution in women's education and employment rights" in Margaret W. Rossiter, *Women Scientists in America: Before Affirmative Action 1940–1972* (Baltimore, MD: Johns Hopkins Press, 1995): 382.

[4] Library of Congress, THOMAS, S-568; H.R.5305: Women in Science and Technology Equal Opportunity Act, http://thomas.loc.gov/cgi-bin/bdquery/z?d096:SN00568:@@@L.

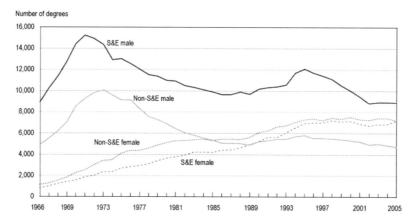

FIGURE 2.3: Doctoral degrees awarded in S&E and non-S&E fields to U.S. citizens and permanent residents by sex: 1966–2005. Source: National Science Foundation, Division of Science Resources Statistics, Survey of Earned Doctorates, 1966–2005, http://www.nsf.gov/statistics/wmpd/figf-1.htm.

While Title IX often receives the most attention for its affiliation with athletics, taken together these legislative acts established an overarching education and employment policy embedded in a legal infrastructure with implications for women in S&E. Part and parcel of this initiative was, for instance, the creation of provisions in three lasting areas of interventions—in "public understanding," educational institutions, and equal opportunity employment—all of which were to be facilitated by the NSF. In many respects, this legal infrastructure—especially in its stipulated enforcement and compliance procedures—is an underutilized instrument to remedy women's underrepresentation in S&E education and employment. As the title of a U.S. Government Accountability Office (GAO) July 2004 Report indicates, *Women's Participation in the Sciences has Increased, but Agencies Need to Do More to Ensure Compliance with Title IX.*[5] Although "[d]iscrimination against women in areas such as college admissions, intercollegiate athletics, and employment was widespread 40 years ago," it notes, and though "civil rights laws in the 1960s barred discrimination in employment," it took Title IX of the Education Amendments of 1972 to "extend" these protections "to students and faculty" by "prohibiting discrimination on the basis of sex in education programs and activities receiving any federal financial assistance" (GAO 2004, 1).

But while research shows that Title IX "has contributed to greater inclusion of women and girls in sports programs," there is "little awareness that the law applies to academics," and, further,

---

[5] July 2004 Report to Congressional Requesters, Gender Issues: *Women's Participation in the Sciences has Increased, but Agencies Need to Do More to Ensure Compliance with Title IX* (GAO-04-639) (U.S. Government Accountability Office, Washington, D.C.), http://www.gao.gov/new.items/d04639.pdf. For NSF engagement with the substance of this report, see http://www.nsf.gov/od/oeo/freq_questions.pdf.

that in an area which receives "billions of dollars in federal assistance," there are problems of compliance with Title IX (GAO 2004, 1, 11). The issue is not only whether federally funded programs are complying with the law, but whether federal science agencies are performing their monitoring role. For instance, grantees must set up compliance procedures, inform all participants (students, employees) that discrimination is illegal in their programs, demonstrate how their programs follow Title IX's rules, and assign a coordinator for these efforts. The reality is that, as of 2004, few mechanisms were in place for the Department of Education, the Department of Energy, the National Aeronautics and Space Administration (NASA), and the NSF to review or ensure compliance by federal grant recipients with Title IX, and "[m]ost have not conducted all required monitoring activities"—all of which leaves federal and participating academic institutions open to legal suit (GAO 2004, 1).

With the advent of this policy, billions of dollars have been invested toward these ends—notably by NSF in its congressionally mandated role to oversee and support activities "designed to increase the participation of women" in STEM fields, but also by other federal science agencies. These investments are accompanied by sustained national research into this critical need area, including continuous tracking of data (also required by law). But one of the most important outcomes has been the fostering of a national conversation about the problem of "low numbers of women in S&E" which, in turn, has helped to aid change—in both the academic climate and the culture of employment in science and engineering at large.[6]

One indication of the force of this conversation—as well as the critical consciousness it has helped to form—is in the backlash against former Harvard President Lawrence Summers' speculative remarks on women's aptitude in science. Indeed, his comments sparked critiques by university presidents and leaders, a spate of news articles, issue-oriented websites and blogs, and scholarly articles reviewing the research.[7] As many of these commentators asserted, "[b]iological explanations" for the low number of women professors in science and engineering "have not been confirmed by the preponderance of research" (whether in brain studies, hormones and performance, cognitive devel-

---

[6] NSF Division of Science Resources Statistics (SRS) fulfills the legislative mandate to "provide a central clearinghouse for the collection, interpretation, and analysis of data on scientific and engineering resources, and to provide a source of information for policy formulation by other agencies of the Federal Government." The overview series, issued since 1982, "Women, Minorities, and Persons with Disabilities in S&E," documents short and long-term trends in the participation of women, minorities, and persons with disabilities in S&E fields across such topics as higher-education enrollment, degrees, financial support, and employment status, including sectors and salaries. See NSF 2007 [246].

[7] The Women in Science & Engineering Leadership Institute (WISELI) University of Wisconsin-Madison, "Responses to Lawrence Summers on Women in Science," http://wiseli.engr.wisc.edu/news/Summers.htm; see also John Hennessey et al. 2005 [171]

opment, etc.), and, in fact, the "dramatic increase in the number of women science and engineering PhDs over the last 30 years clearly refutes long-standing myths that women innately or inherently lack the qualities needed for success," since "obviously, no changes in innate abilities could occur in so short a time."[8]

We will revisit these three critical interventions—in public discourse, educational institutions, and employment. In many ways, they inform the program and research agendas for women in science and engineering at many colleges and universities which appeal to these federally mandated "areas of intervention" to explain, defend, and promote institutional change. Suffice it to say, here, however, that while important gains have been made in each area—most notably in public awareness of underrepresented groups in S&E as part of a significant national and international problem, and in the importance of increasing women's participation and representation in educational institutions—the third area, employment, including academic employment, still requires critical and creative thinking and resource investment to achieve similar transformations.

2.3   REMAINING CHALLENGES

While it is certainly true that powerful efforts by government, private industry, and educational institutions to redress the gap of women, minorities, and the disabled in S&E fields has reaped important results and, thus, should be treated as a tentative success story, there are caveats that qualify this success. We briefly relay five persistent challenges in Table 2.1 that define the contemporary "field of struggle" for women in S&E with implications for women leaders in the academic context. We should also say that, although we focus our attention on women faculty and academic leaders in S&E, their insights offer broader lessons for women across different positions of leadership more broadly.

2.3.1   Gendered Field Segregation and Advancement

The first challenge involves the limited usefulness of numbers—what we call the ruse of representation—in understanding this critical need area and in adjudicating genuine change or institutional transformation. While women's increased participation in S&E is a necessary and inspiring first step,

---

[8] J. Handelsman et al., "More Women in Science," 1190; *Beyond Bias and Barriers*, 215. See also D. Halpern, C. Benbow, D. Geary, R. Gur, J. Shibley Hyde, and M.A. Gernsbacher, "The Science of Sex Differences in Science and Mathematics," *Psychological Science in the Public Interest*, 8.1 (2007) 1–51; Elizabeth Spelke, "Sex Differences in Intrinsic Aptitude for Mathematics and Science?: A Critical Review," *American Psychologist*, 60.9 (Dec 2005): 950–958; Cody Ding, Kin Sing, and Lloyd Richardson, "Do Mathematical Gender Differences Continue? A Longitudinal Study of Gender Difference and Excellence in Mathematics Performance in the U.S.," *Educational Studies: Journal of the American Educational Studies Association*, 40.3 (2007): 279–295; *The Edge*, "The Science of Gender and Science, the Pinker vs. Spelke Debate" (16 May 2005), www.edge.org/3rd_culture/debate05/debate05_index.html.

TABLE 2.1: Five challenges defining the "field of struggle" for women in S&E	
1	Gendered field segregation and advancement
2	Disaggregating "diversity"
3	Transforming S&E climates (not integrating women into existing hierarchies)
4	Fully utilizing the pool of qualified, talented S&E candidates
5	Leadership: Promoting women's authority in S&E education

progress belongs disproportionately to education. By contrast, while women make up half of the total U.S. workforce, they comprise only one fifth of science, engineering, and technical workers.

Moreover, in the educational arena there are two areas in which these statistical "wins" reveal troubling, persistent gendered issues: (1) gendered field segregations evident in the low percentage of women earning degrees in engineering, the physical sciences, and computer science that continue into the professional-career context; and (2) the discrepancy between the pipeline, the available pool of up-and-coming women S&E doctorates, and women in advanced professional S&E careers, especially at the upper-echelons of academic institutions.[9] In the first case, while women in 2002 earned more than half of all bachelor's degrees, including degrees in S&E, significant variations persist among these fields. Women, for instance, earned the following percentage of bachelor's degrees in psychology (78), biological/agricultural sciences (59), social sciences (55), mathematics (47), engineering (21), computer sciences (27), and physical sciences (43).[10] Such discrepancies indicate the variegated nature, the different cultures of science and engineering—the increasing prevalence and acceptability, for instance, of women physicians versus women physicists—both at the level of a given institution and in our culture at large.[11]

The second aspect of this problem arises from gendered obstacles in advancement. The discrepancy between the available pool of women in the pipeline and the numbers of advanced women

[9] Donna J. Nelson, 2005, *A National Analysis of Diversity in Science and Engineering Faculties at Research Universities*, Norman, OK (revised Oct 2007), http://cheminfo.chem.ou.edu/~djn/diversity/briefings/Diversity%20Report%20 Final.pdf.

[10] National Science Board, 2006, *Science and Engineering Indicators, 2006*, (NSB 06-02) Arlington, VA: NSF, 2-5, http://www.nsf.gov/statistics/seind06/c2/c2h.htm.

[11] NAS, *Beyond Bias and Barriers* (2007, 15) found, when comparing the PhD pool with professor positions (in 2003), that declines were larger in fields that required postdoctoral study (i.e., life sciences, chemistry, and mathematics), whereas fields that had an initial low number of women in the undergraduate/graduate pool (i.e., computer science, physical sciences) retained a more constant proportion of women at the faculty rank.

scientists and engineers in positions to train and mentor the next generation (illustrated in Table 1.1) appears, at first glance, to be a clear-cut matter of numbers. Hence, the solution was to expand the pipeline to, thereby, increase representation. In fact, early studies discovered in support of this case that matching the gender of faculty to students fostered a positive attitude toward science.[12]

But over the last few years, the picture has been complicated by new studies which have found that the "problem is not simply the pipeline," since, while many fields now have gender parity (NAS 2007, 2), women still only comprise 15.4% of full professors in the social/behavioral sciences and 14.8% in the life sciences at top research universities—the "only fields" in S&E where "the proportion of women reaches into the double digits." Lack of female mentors for S&E students to identify with is not the only issue. Superb candidates are not often encouraged to pursue S&E careers; advisers may be unaware of the barriers that women face, or feel unsure about how to mentor women in light of these barriers; and few preparation programs may be available to help students transition to faculty (i.e., providing access to role models, or inspiring confidence and commitment), such as those offered by many professional societies, universities, and private organizations.[13] In many cases, then, "closing the gap" requires more systemic and intentional strategies and programs, "explicit" and straightforward "encouragement of outstanding doctoral candidates to enter the professoriate" in an organized and consistent fashion, and dealing with and redressing problems of institutional culture from the get go in order to translate increased numbers of women at the undergraduate and graduate level into faculty and ultimately leadership positions (Handelsman et al. 2005, 1191).

At the heart of this challenge is the need to shift from an exclusive focus on numbers to investigations of culture. Indeed, practitioners in the academic context are beginning to find, not only that the discrepancy between women S&E students and faculty is attributed to multicausal factors—unintentional biases, outdated institutional policies and practices, evaluation bias, family support systems, etc.—but that these factors may be subtle, implicit, and systemic, not the product of any individual prejudice or ill-will, and, thus, difficult to identify.[14]

---

[12] Walter S. Smith and Thomas Owen Erb, "Effect of Women Science Career Role Models on Early Adolescents' Attitudes Toward Scientists and Women in Science," *Journal of Research in Science Teaching* 23.8 (Nov 1986): 667–676.

[13] M.F. Fox, "Gender Faculty and Doctoral Education in Science and Engineering" and Charlotte Kuh, "You've Come a Long Way: Data on Women Doctoral Scientists and Engineers in Research Universities," in *Equal Rites, Unequal Outcomes: Women in American Research Universities*, ed., L. S. Hornig (NY: Kluwer Academic, 2003) 91–109 and 111–144.

[14] Y. Xie and K.A. Shauman, *Women in Science: Career Processes and Outcomes* (Cambridge, MA: Harvard University Press, 2003); Virginia Valian, *Why So Slow? The Advancement of Women* (Cambridge, MA: MIT Press, 1999); Henry Etzkowitz, Carol Kemelgor, and Brian Uzzi, *Athena Unbound: The Advancement of Women in Science and Technology* (Cambridge, UK: Cambridge University Press, 2000); Sue Rosser, *The Science Glass Ceiling: Academic Women Scientists and the Struggle to Succeed* (NY: Routledge, 2004).

2.3.2 Disaggregating "Diversity"

The second impediment to women's full participation in S&E fields is a general misunderstanding of diversity, including narrow or superficial definitions of difference which do not take into account or leverage the value of diversity in the S&E enterprise. While the NSF, especially in its ongoing reports, *Women, Minorities, and Persons with Disabilities in Science and Engineering*, has spearheaded attending to gender, as well as ethnicity/race, disability, and even foreign/international status, there is still little in-depth understanding and qualitative research on the range of perspectives, experiences, and needs that comprise the umbrella terms "women" or "minority" in S&E and, thus, what diverse constituencies bring to these fields. We know little about the range of experiences among women academic scientists and engineers, how the experiences of women of color are internally variegated, for instance, or how they differ from white women's experiences—not to mention how these differences inform styles of leadership. One recent study, for instance, noted different expectations among African-American and white women engineering undergraduate majors toward dominant stereotypes of the engineer as presumptively male/masculine[15] (Hanson 2004).

Likewise, the representational issue changes when viewed through the lens of race/ethnicity and nationality. For instance, although African-Americans and Latinos are less likely than whites and Asians to graduate from high school or to enroll in and graduate from college, those who do choose S&E majors about as often as whites do. Asians, by contrast, were much more likely to choose S&E fields: almost half of all bachelor's degrees earned by Asians are in these fields, as compared to about one third for the other racial/ethnic groups. Although these examples are drawn from high school and undergraduate students, they imply circumspection in aggregating all experiences under simplistic racial, ethnic, or gender terms (i.e., "woman" or "minority"), and caution in making generalizations about these experiences by virtue of group belonging.

Attending to the complexity of identity—including how multiple identity positions may overlap and intersect with each other (i.e., woman, international student, person of color)—allows for deeper questioning and analysis in present attempts to diversify academic leadership, such as what accounts for differences among identity groups, or what patterns defy or confirm cultural expectations and stereotypes in S&E. Most important, a more critical attitude toward diversity allows

---

[15] Sandra L. Hanson. "African American Women in Science: Experiences from High School through the Post-Secondary Years and Beyond," *NWSA Journal* 16.1 (2004) 96-115; Josephine Beoku-Betts, "African Women Pursuing Graduate Studies in the Sciences: Racism, Gender Bias, and Third World Marginality," *NWSA Journal* 16.1 (2004) 116–135; Sandra Harding, *Is Science Multi-Cultural? Postcolonialisms, Feminisms, and Epistemologies* (Bloomington: Indiana University Press, 1998); Patricia Hill Collins, "Moving Beyond Gender: Intersectionality and Scientific Knowledge," in eds., Myra Marx Ferree, Judith Lorber, and Beth B. Hess, *Revisioning Gender* (Thousand Oaks, CA: Sage, 1999): 261–284.

for a continual updating and redefining of our understanding of diversity, not only in a multicultural society, but for enabling a more accurate appreciation of the range of perspectives and experiences that comprise the S&E fields and a more creative approach to the range of solutions needed for today's S&E problems, including increasing women's leadership prospects.

The second aspect of the diversity challenge stems from reductive or conflicting definitions of "minority" and more hotly contested terms such as "race" and "ethnicity," and a narrowing of the field of diversity so that such variables as nationality, culture and background, academic discipline, career trajectory, and, most important, leadership style are overlooked. We briefly expand upon this issue by emphasizing the range of experiences and perspectives that comprise diversity, as well as a critical approach to language in discussing diversity in relation to women in science and engineering. The NSF, for instance, distinguishes "minority"—referring to all ethnic and racial groups "other than white"—from the term "underrepresented minority," which is used to designate, more literally, the three groups whose numerical representation in S&E amounts to less than their percentage in the general U.S. population: i.e., African Americans, Hispanics, and American Indian/Alaska Natives.[16] While the term "underrepresented minority" hones in on ethnic identity in relation to numerical representation in the S&E fields, the term "minority" itself is fraught for many reasons. In this specific context, for instance, designated minorities may actually be majorities in some situations (i.e., certain states, cities, programs, universities/colleges, employment contexts), and the term does not specify what constitutes a non-minority (i.e., whiteness), or even how persons from mixed parentage should be designated.[17] While Asians as well as African-Americans, Latinos, and Native Americans are traditionally defined as ethnic "minorities" in U.S. culture, for instance, only the latter would count as "underrepresented minorities" in their numerical representation in S&E.

In all cases, however, the use of "minority" leaves one with the impression that numbers are the problem, even while this exclusive focus, as we have suggested, and the use of a terminology (i.e., minority) beholden to numerical representation, cannot tell the full story, nor illuminate the multiple paths for institutional change. Women, for instance, now outnumber men in undergraduate education (since 1982) and earn at least half of all S&E bachelor's degrees (since 2000), which makes concerns strictly for their numerical "underrepresentation" fraught. Indeed, if concerns over representational parity were enough, we would not have a discrepancy from pipeline to profession.

---

[16] NSB *Science and Engineering Indicators 2006*, 2–37, www.nsf.gov/statistics/seind06/c2/c2g.htm#underrepresented.

[17] This usage, for instance, can be seen in the distinction between Asians and other minorities in the June 2006 *InfoBrief Academic Institutions of Minority Faculty with S&E Doctorates*, which notes: "Although the number of S&E doctorates awarded to Asians is much larger than the numbers awarded to blacks, Hispanics, and American Indians, most (85 percent) are non-U.S. citizens on permanent resident or temporary visas." National Science Foundation, Division of Science Resources Statistics, Joan S. Burrelli, InfoBrief, *Academic Institutions of Minority Faculty with S&E Doctorates* (NSF 06-318) Arlington, VA , June 2006, 1, www.nsf.gov/statistics/infbrief/nsf06318/nsf06318.pdf.

Such concerns, additionally do not take into consideration how women, even when they achieve significant numbers or representational parity with men, still face gender-based discrimination and biases, or why women leave S&E in greater numbers, despite their success. In medical schools, for instance, women in 2004–2005 outpaced men as a percentage of applicants (50.4), and came close to achieving numerical parity with matriculated men (49.5) and with graduates (47.1). Yet, the occurrence of gender discrimination and sexual harassment had not been reported to have changed in medical education or the professional workplace, according to a survey of 214 physician mothers and their physician daughters.[18] Indeed, it concluded that "Leadership of medical institutions and professional associations must deal more effectively with persistent discrimination and harassment or risk the loss of future leaders" (Shrier et al. 2007, 883). Not only will numbers alone, then, not solve the problem, they do not help explain how in certain cases groups with small numbers do not face bias or discrimination, such as male nurses. We must, therefore, shift our conceptual lexicon from an exclusive focus on diversity as a matter of representation to diversity as involving power relations—explicit and implicit biases, discrimination and barriers, questions of social roles and cultural norms.

The key criterion for promoting diversity in a thoughtful manner, then, is not an exclusive focus on numbers (though numbers are important), but a consideration of "historical subordination in relation to specific ethnic and racial hierarchical systems of value and status, and/or non-dominant positions with respect to a given state."[19] In this sense, alternative terms, such as "women of color" instead of "minority women," may help to challenge the implication that numbers alone are the determining issue.[20] Therefore, while we also use the term "underrepresented minority" as defined by the NSF for consistency's sake, we wish to emphasize the limitations in focusing exclusively on numbers, or on genealogical origins and regional place of birth, for instance, as opposed to "patterns of persistent discrimination," when defining certain identity groups in S&E.

Perhaps the most pernicious side effect of focusing exclusively on numbers—and in defining categories of difference in those terms—is the inability to see the specific benefits of diversity. We therefore emphasize, along with a group of recent scholarship, the educational and cognitive benefits of diversity, how the heterogeneity of students, faculty, and staff can strengthen colleges/

[18] Association of American Medical Colleges (AAMC) *Women in Academic Medicine Statistics and Medical School Benchmarking* 2004–2005.

[19] The United Nations offers a bit more substantive and expanded definition of the term 'minority': "A group of citizens of a State, constituting a numerical minority and in a non-dominant position in that State, endowed with ethnic, religious or linguistic characteristics which differ from those of the majority of the population, having a sense of solidarity with one another, motivated, if only implicitly, by a collective will to survive and whose aim is to achieve equality with the majority in fact and in law," *Proposal Concerning a Definition of the Term 'Minority,'* UN Document E/CN.4/Sub.2/1985/31 (1985).

[20] Syracuse University, *Senate Committee on Diversity, Senate Committee on Diversity Recommendations in Faculty Retention* (Feb 2007), http://universitysenate.syr.edu/diversity/diversity-21mar07.pdf.

universities (i.e., multiple perspectives can enable more innovative solutions to problems, bring a higher level of critical analysis to decision-making, offer more favorable working environments for all members), and how the manifestly diverse fields of S&E may possess an edge in these respects.[21] Indeed, such benefits multiply if one considers such issues as culture, background, academic field, career trajectory, and leadership style—factors which also fundamentally expand the range of solutions for increasing women's leadership prospects in S&E.

2.3.3   Transforming S&E Climates

The third challenge for women academic scientist and engineers arises from a shift in thinking about solutions to the lack of women in the S&E fields: from integrating women into existing S&E climates to transforming them. As Jill Bystydzienski notes, "[u]ntil recently," most "intervention programs" at colleges/universities, which were often supported by government and private foundations, "focused on how to fit women into existing science and engineering departments, programs, and laboratories" because "[i]t was assumed that women were 'deficient' in math and science achievement and lacked motivation to participate."[22] Hence, "they had to be individually encouraged, mentored, supported, and appropriately socialized to enter and remain in the sciences, engineering, or technology fields" (2004, ix). Yet, Bystydzienski goes on to note, the percentages of women and other underrepresented groups in most S&E fields "remain alarmingly small," even though "the gaps between the sexes in achievement and course taking in math and science have become statistically insignificant, and performance levels, for example, of engineering women faculty and faculty of color are no different from those of white men" (2004, ix). The result is that:

> Activists and policymakers have thus increasingly recognized what many women's studies scholars have been advocating for some time now: That the (remaining) barriers to women's progress in academia are systemic and rather than trying to change women to fit the sciences and engineering, these fields need to be changed in order to accommodate women (2004, ix).

Instead of attempting to "fit" women into preexisting curricula and climates, scholars in S&E, the social sciences, in women's studies/feminist, in psychology, education, and public policy, often in

---

[21] For research on the educational value of diversity, see Gurin et al. 2002 [154] Gurin et al. 2004 [155]; Hurtado 2007 [180]; Milem 2003 [233]; Milem et al. 2005 [234].

[22] Jill Bystydzienski, "(Re)Gendering Science Fields: Transforming Academic Science and Engineering," *NWSA Journal* 16.1 (2004) viii–xii, ix. See also B.C. Clewell and P.B. Campbell, "Taking Stock: Where We've Been, Where We Are, Where We're Going," *Journal of Women and Minorities in Science and Engineering* 8.3&4(2002) 255–284; Judy Jackson "The Story Is Not in the Numbers: Academic Socialization and Diversifying the Faculty," *NWSA Journal* 16.1 (2004) 172–185.

dialogue with women in S&E program facilitators, have helped to develop a theoretical framework and critical language to initiate creative innovation in S&E research and pedagogy.

One profound outcome of this interdisciplinary effort is the felt need to rethink equity and equal opportunity, not only as matters of normative commitment in higher education, but as fundamental goals in achieving educational and scientific quality, excellence, and progress.[23] Interest in multicultural science and engineering, social and global responsibility, and new programmatic recommendations (i.e., learning communities, online advising, mentoring networks, women in science and engineering or WISE programs etc.) all are designed to make the college/university setting more accountable to and inclusive of underrepresented groups.[24] These efforts have also interfaced with other calls to innovate the S&E fields to meet new and imminent demographic realities, real-world challenges (i.e., globalization, technological changes, emerging subfields), an interest in the role of technology in social justice and environmental sustainability, and the importance of interdisciplinary collaborations across S&E and other fields.[25] Such issues raised by these efforts include: retaining a U.S. global competitive edge and U.S. science–technology leadership globally; exploring the social justice aspects of S&E; cultivating critical analytical and cultural-sensitivity skills in

---

[23] P.L. McLeod, S.A. Lobel, and T.H. Cox, "Ethnic Diversity and Creativity in Small Groups," *Small Group Research* 27 (1996) 248–265; Jeffrey Milem, "The Educational Benefits of Diversity: Evidence from Multiple Sectors," in *Compelling Interest: Examining the Evidence on Racial Dynamics in Colleges and Universities*, eds., Mitchell J. Chang et al. (Stanford, CA: Stanford University Press, 2003) 126–169; Charlan J. Nemeth, "Dissent as Driving Cognition, Attitudes, and Judgments," *Social Cognition* 13 (1995): 273–291; Daryl Smith, et al., Diversity Works: *The Emerging Picture of How Students Benefit* (Association of American Colleges and Universities, Washington D.C., 1997). For a technical approach to this issue, see Scott E. Page, *The Difference: How the Power of Diversity Creates Better Groups, Firms, Schools, and Societies* (Princeton, NJ: Princeton University Press, 2007) and Lu Hong and Scott E. Page, "Groups of Diverse Problem Solvers Can Outperform Groups of High-ability Problem Solvers," Proceedings of the National Academy of Sciences/PNAS, 101(46), 16 Nov 2004: 16385–16389.

[24] See Sandra Harding, *Is Science Multicultural? Postcolonialisms, Feminisms and Epistemologies* (Blooming, IN: Indiana University Press, 1998). An interesting example of this changing climate of S&E is the explosion of "hands-on" programs, such as Engineers without Borders, in which participants use technical expertise to solve social or global problems and "deliver sustainable solutions" to improve conditions for disadvantaged communities worldwide. See Engineers without Borders-USA (EWB-USA), 2008, http://www.ewb-usa.org.

[25] Donna Riley and Caroline Baillie, *Engineering and Social Justice* (Morgan & Claypool Publishers, 2008); Caroline Baillie, *Engineers within a Local and Global Society* (Morgan & Claypool Publishers, 2006); R.M. Felder, S.S. Sheppard, and K.A. Smith "A New Journal for a Field in Transition," *Journal of Engineering Education* 94.1 (Jan 2005): 7–9; S.K. Bhatia and Jennifer Smith, *Bridging the Gap between Engineering and the Global World: A Case Study in the Coconut (Coir) Fiber Industry in Kerala, India* (Morgan & Claypool Publishers, 2008); D.A. Vallero and P.A.Vesilind, Socially Responsible Engineering (John Wiley & Sons, 2007); Shirley M. Tilghman, "Engineering for a Better World: A Vision for Princeton," 12 May 2004, http://www.princeton.edu/president/pages/20040512/index .xml. According to the recent National Academies Report (NAS/NAE/IOM: *Rising above the Gathering Storm: Energizing and Employing America for a Brighter Economic Future*, Washington, DC: The National Academies Press, 2007), S&E education and research are "increasingly global endeavors" and globalization has "begun to challenge the longstanding scientific pre-eminence of the United States and, therefore, its economic leadership."

students; broadening the vision of what it means to be a scientist and engineer or what counts as science/engineering problems and solutions; and the role of the university in teaching these insights as part of the curricular content.

2.3.4   Fully Utilizing the Pool of Qualified, Talented S&E Candidates

Fourth, the National Academies' *Beyond Bias and Barriers, Fulfilling the Potential of Women in Academic Science and Engineering* (2007) argues that "the pipeline"—increasing the available pool of qualified women graduating with S&E degrees—is no longer "the problem," as mentioned, but *fully utilizing* this pool of qualified candidates defines the newest challenge. We have noted, for instance, the serious disparity between women earning Ph.D.s and those in faculty or leadership positions at top research universities.[26] This challenge requires a different approach, such as tackling institutional and cultural barriers, unconscious bias and accumulated disadvantage—climate factors which make all the difference between, not only whether women persist, but how they thrive. The problems of "squandered talent" and "competing in a disparaging climate," for instance, have been recently publicized, in part, because of the entrenched misunderstanding about women in S&E fields held even by leading academic stakeholders and gatekeepers, such as Lawrence Summers—including persistent stereotypical thinking about women scientists and engineers.[27] Such assumptions also embed themselves in university and department policies and practices, search and hiring committees, candidate evaluations, professional support, promotion, and advancement, and performance evaluations. They occur in informal practices, such as, whether or not, for instance, women develop variegated support networks to help craft their own career paths and advancement.

2.3.5   Leadership: Promoting Women's Authority in S&E Education

All of these challenges—pressing for quality (not only quantity) in academic life, leveraging diversity as a matter of excellence, innovating the S&E fields, fully utilizing available technical talent—help to clarify the fifth and final challenge and the priority of this book: academic leadership for women in S&E. Indeed, leadership is that area that still needs significant critical attention.

Recalling the tripartite areas of federal intervention set out in the *Science and Engineering Equal Opportunities Act*, there is relative success, as mentioned, in two of the three areas—higher education and public understanding. Indeed, positive results are evident in light of this legislation's

---

[26] Niemeier and González (2004).

[27] Marcella Bombardieri, "Summers' Remarks on Women Draw Fire," *The Boston Globe*, 17 Jan 2005; Cornelia Dean, "Women in Science: The Battle Moves to the Trenches," *New York Times*, 19 Dec 2006 and "Bias Is Hurting Women in Science, Panel Reports," *New York Times*, 19 Sept 2006; John Hennessey, Susan Hockfield and Shirley Tilghman "Women in Science: The Real Issue," *Boston Globe*, 12 Feb 2005. See also the collection of "Responses to Lawrence Summers on Women in Science" http://wiseli.engr.wisc.edu/news/Summers.htm.

goals to (1) increase "the participation of women in scientific and technical studies, training, fellow-ships, and careers" and (2) develop a "clearinghouse" and research programs to promote "the partici-pation of women in science and technology," educate the public on "the potential contribution of women in these fields," and identify and develop "books and instructional materials" towards these ends. The statute's third criterion, however, Title III on equal opportunity employment, represents a more demanding and neglected area—and the most legally binding.[28] It explicitly requires, for in-stance, that "the head of each Federal agency, national laboratory, and federally-funded research and development center in science and technology: (a) prevent discrimination against women in science and technology; (b) increase opportunities for the employment and advancement of women in these fields; and (c) encourage the participation of minority and physically handicapped women in science and technology careers." At every level, leadership is critical to achieving these requirements.

In fact, judging from the perspective of this third federally mandated priority of equal oppor-tunity employment and increased employment opportunities, the very data, the numbers, that once seemed so promising, may appear inadequate. As Judy Jackson has observed, "the story" of women's participation in the S&E is not one that can be told by "the numbers" alone.[29] In fact, a reliance on quantitative data has often been achieved at the expense of qualitative inquiry into women leaders' experiences, perspectives, and perceptions. In addition, evaluative studies have often focused on single institutions, especially Research 1 institutions, and emphasized personal-anecdotal informa-tion from faculty in these contexts,[30] while interdisciplinary research on gender and leadership from other perspectives (i.e., business-management and the public sector) are often not incorporated into these assessments. Without engaging a broad range of research and inquiry, it is difficult to under-stand the nature and meaning of the changes in numbers that have occurred thus far, or to enhance the outcomes for all women. Indeed, without seriously contending with the challenges that underlie the numbers, women faculty coming up through the ranks may think twice about undertaking the additional scrutiny and pressure that leadership, especially for women, is known to bring.[31]

---

[28] For 42 USC 1885, (a) and (b), see http://rand.org/scitech/stpi/AuthAct.pdf.

[29] Judy Jackson 2004 [186], "The Story Is Not in the Numbers: Academic Socialization and Diversifying the Fac-ulty," *NWSA Journal* 16.1 (2004) 172–185; Susan Madsen 2007 [223], "Women University Presidents: Career Paths and Educational Backgrounds," *Academic Leadership: The Online Journal* 5.1 (2007); B. Avolio, et al 2003 [22], "Leadership Models, Methods, and Applications," in *Handbook of Psychology*, eds. D. K. Freedheim et al. (Hoboken, NJ: Wiley: 2003): 277–307: 287.

[30] "Nelson Diversity Surveys" 2005; Massachusetts Institute of Technology (MIT), "A Study on the Status of Women Faculty in Science at MIT (Cambridge, MA: MIT, 1999).

[31] A.H. Eagly and S.J. Karau, "Role Congruity Theory of Prejudice Toward Female Leaders," *Psychological Review* 109.3(2002): 573–598; A.H. Eagly and L.L. Carli, "The Female Leadership Advantage: An Evaluation of The Evi-dence," *The Leadership Quarterly* 14 (2003): 807–834; M. Heilman and T.G. Okimoto, "Why Are Women Penalized For Success at Male Tasks?: The Implied Communality Deficit," *Journal of Applied Psychology* 92.1(2007): 81–92.

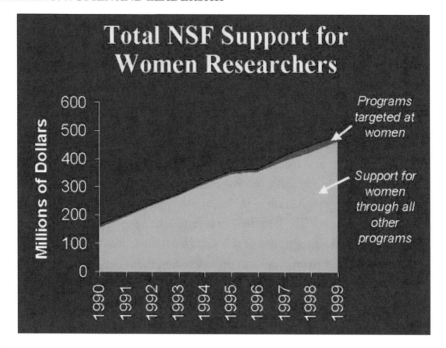

FIGURE 2.4: Rita R. Colwell, Director, National Science Foundation, 1998–2004 (2001)[32]

Skepticism toward a focus on numbers might also be warranted in considering other arenas used to mark women's progress in S&E. According to former NSF Director Rita Colwell's comments, for instance, NSF programs that specifically target women (i.e., ADVANCE, among others) represent a small percentage of overall funds available to women, since other NSF grants actually amount to a far greater portion of NSF's total annual budget (see Figure 2.4). A critical question would be whether these funds for women *outside* targeted programs enhance the advancement or success of women researchers, and, further, whether these additional and combined funds are proportionate to the number of participating women. As Table 2.2 shows, programs designed to aid in "broadening participation" for underrepresented groups represent only a small portion of NSF's total annual budget. If the NSF budget in 1999 was 3.672 billion, for instance, and support for women in targeted and non-targeted programs was between 450 and 500 million, which represents

[32] Rita R. Colwell, "Barriers to and Opportunities for Women in Science," Speech to Washington College (Arlington, Virginia: National Science Foundation, 2001), www.nsf.gov/news/speeches/colwell/rc011017washcollege.htm.

PROGRAM	EXAMPLES (NOT COMPREHENSIVE)	2006	2007	2008 REQUEST
Programs for minority individuals	AGEP, LSAMP	72.32	82.80	86.19
Programs for minority institutions	CREST, HBCU-UP	62.42	76.85	83.96
Gender-based programs	ADVANCE, GSE	37.34	38.78	38.59
programs for persons with disabilities	RDE	5.30	6.00	6.00
Other Broadening Participation Programs	ATE, H-1B, ISE	443.05	436.91	443.98
Total		620.43	641.34	658.72
Total NSF (billions)		5,581.17	5,917.16	6.43

TABLE 2.2: NSF programs to broaden participation budget, 2006–2008

approximately 12–15% of the budget going to women recipients for this year—the question would be whether this number is adequate or modest compared to the total number of women scientists and engineers that the federal government supports.

Critically probing surface representations of success is an ongoing responsibility that reveals a more complex picture—one that invariably involves the challenge of leadership.

CHAPTER 3

Gender and Leadership: Theories and Applications

3.1 INTRODUCTION

Interdisciplinary research on gender and leadership is often not incorporated into understanding the nature and meaning of women faculty experiences in S&E. We, thus, provide a context for the changes that have occurred for women academics, including the theoretical frameworks that are likely to aid and enhance future outcomes.

3.2 LEADERSHIP TYPOLOGIES

Real institutional change is fundamentally dependent upon strong and supportive leaders.[1] *Beyond Bias and Barriers* (2007, 219) notes, for instance, since "[c]areer impediments for women deprive the nation of an important source of talented and accomplished scientists and engineers," transforming institutional "structures and processes to eliminate gender bias requires a major national effort, incorporating strong leadership and continuous attention, evaluation, and accountability." The use of leadership for institutional transformation—where leaders are seen as "vital in achieving organizational objectives or to instigate organizational change"—is often termed the "new public management" in Europe, and represents a moment when a key observation from organizational theory has become common sense among higher-education stakeholders.[2]

[1] NAS, *Beyond Bias and Barriers* 2007: 7; National Research Council, *To Recruit and Advance: Women Students and Faculty in U.S. Science and Engineering,* (Washington, D.C.: National Academies Press, 2006); The Best Initiative, *The Talent Imperative: Diversifying America's Science and Engineering Workforce,* (San Diego, CA, 2004b), www.bestworkforce.org/PDFdocs/BESTTalentImperativeFINAL.pdf: 5; The Best Initiative, *A Bridge for All: Higher Education Design Principles to Broaden Participation in Science, Technology, Engineering and Mathematics* (San Diego, CA, 2004a), www.bestworkforce.org/PDFdocs/BEST_BridgeforAll_HighEdDesignPrincipals.pdf: 23.

[2] Berit Askling and Bjørn Stensaker, "Academic Leadership: Prescriptions, Practices and Paradoxes," *Tertiary Education and Management* 8 (2002): 113–125: 113; T. Christensen and P. Lægreid, "A Transformative Perspective on

But leadership, especially when women are involved, is not only an instrumental issue, a tool for running organizations better: it involves matters of authority, as well as change. Understanding why requires a brief discussion of leadership itself and, more to the point, an appreciation of how, as many in the interdisciplinary field of leadership studies attest, leadership is a notoriously "complex, fragmented and contradictory" endeavor and, as such, the subject of much debate.[3] As Joseph Rost in *Leadership for the Twenty-first Century* (1991, 92, 149) points out, not only are the words that scholars use to define leadership "contradictory," its models "discrepant," including a "confusion of leadership with management" and the equation of leaders with leadership, these and other "serious conceptual problems" are "hard to reconcile" with "the real world" of leadership. The very idea of the leader has been "exploited," he continues, for purposes of "symbolic mythmaking"—the notion of the leader as a mythical savior of organizations—an indication of how much "the concept has lost its moorings, if not its essential character" (1991, 92).[4]

Likewise, the very premise that "good leadership" can be learned, studied, and taught, invariably by a cohort of expert "leadership scholars, teachers, consultants, trainers, and coaches," notes Barbara Kellerman, is also a very recent and debated idea—and one that has "become big business," a "leadership industry," which includes over 600 leadership development programs at U.S. post-secondary institutions in a number that has more than doubled in the last four years.[5] Despite longstanding debates, a leadership industry, and ample investments (evident in some business school endowments), fundamental questions about defining leadership still remain as tension points in the field. Is the leader the equivalent of the manager, for instance, or is there something

---

Administrative Reforms," in T. Christensen and P. Lægreid eds., *New Public Management: The Transformation of Ideas and Practice* (Aldershot: Ashgate, 2001a).

[3] Martin M. Chemers, *An Integrative Theory of Leadership* (Mahwah, NJ: Lawrence Erlbaum, 1997): ix. See also Joseph Rost, *Leadership for the Twenty-First Century* (Westport, CT: Praeger, 1991): 13–36; 92; Jeffrey Pfeffer, "The Ambiguity of Leadership," *Academy of Management Review* 2, 1977.

[4] See also Gary Yukl, *Leadership in Organizations* (Upper Saddle River, NJ: Prentice-Hall, 2002); Michael Hoppe, "Cross-Cultural Issues in Leadership Development," in Cynthia McCauley, Russ Moxley and Ellen Van Elsor eds., *The Center for Creative Leadership Handbook of Leadership Development* (San Francisco, CA: Jossey-Bass, 1998).

[5] Barbara Kellerman, *Bad Leadership: What It Is, How It Happens, Why It Matters* (Cambridge, MA: Harvard Business School Press, 2004): 5. If Kellerman conveys leadership and its study as an industry, indeed, one that legitimates the explosion in professional management programs across the country, James Downey, "Guest Editor's Introduction: Academic Leadership and Organizational Change," *Innovative Higher Education* 25.4 (2001), 235–238: 235, notes that Americans "have greater expectations of their leaders," and that "the study of leadership, academic, corporate, religious, and political, is a well-developed industry in the United States."

value-based associated with leading?[6] Is leadership a specialized role, or a social influence process?[7] Does leadership flow from social power and authority, or is it based on personal charisma and influence? Does leadership arise from intentions or practical outcomes, is it the product of visionary principles or effective practices, and is it based on reason or emotion? As Stogdill notes, "there are almost as many definitions of leadership as there are persons who have attempted to define the concept."[8]

Since the dominance of early behavioral theories of the 1970s, the general consensus on how to study leadership has itself undergone change over time. In the 1980s and 1990s, interest in leadership style (i.e., charismatic, transformational and spiritual leadership) became popular, as well as an emphasis on emotions, values, and symbolic behavior, including the leader's role in making meaning.[9] Not only were leaders believed to influence the behavior of others through various means (i.e., sources of power, lines of authority, the nature of the task and desired outcome) as described by Yukl, Guinan and Sottolano (1995), but Yukl (2000; 2002) put forth a classification system describing 11 behavioral influence tactics, such as rational persuasion, inspirational appeal, consultation, and coalition-building.[10]

Accounts today, however, tend to displace earlier theories of leadership based on the leader's ability, behavior, style, or charisma and, instead, define leadership by virtue of the relationships or interactions between those involved in the process, i.e., leaders and followers. In this case, leadership does not inhere in the single individual, or stand as a measure of his or her performance or capacity, but operates as a dynamic, a "collaborative endeavor," a process that involves "interactions" between and among leaders, members, and even outside constituencies.[11] In this view, the context—not only

---

[6] M. Buckingham, "What Great Managers Do," *Harvard Business Review* 83.3, March 2005; Q.N. Huy, "In Praise of Middle Managers, *Harvard Business Review* 79.8 (2001): 73–79; J. Gosling and H. Mintzberg, "The Five Minds of a Manager," *Harvard Business Review* 81.11(2003): 54–63.

[7] B.M. Bass and R.M. Stogdill, *Bass and Stogdill's Handbook of Leadership: Theory, Research, and Managerial Applications* (NY: Free Press, 1990).

[8] R.M. Stogdill, *Handbook of Leadership: A Survey of Theory and Research* (NY: The Free Press, 1974): 259.

[9] Gary Yukl, "An Evaluation of Conceptual Weaknesses in Transformational and Charismatic Leadership Theories," *The Leadership Quarterly* 10.2 (Summer 1999): 285–305, 286; *Leadership in Organizations,* 5th ed. (NJ: Prentice Hall, 2002).

[10] G. Yukl, P.J. Guinan, and D. Sottolano, "Influence Tactics used for Different Objectives with Subordinates, Peers, and Superiors," *Group and Organization Management*, 20 (1995): 272–296; G. Yukl and C.E. Seifert, "Preliminary Validation Research on The Extended Version of the Influence Behavior Questionnaire," Paper presented at the Society for Industrial and Organizational Psychology Annual Conference, 2002.

[11] J. Rost, "Leadership Development in the New Millennium," *The Journal of Leadership Studies*, 1.1 (1993): 91–110.

what the leader does—is equally important to a leader's success. At the very least, then, leadership, a term that continues to evolve, comprises a range of meanings and perceptions, including social and cultural norms, and reflects the diversity of human forms of social organization.[12]

For our purposes, we consider theories of leadership that are useful for improving the prospects for women academic scientists and engineers, especially in promoting institutional change—an area of leadership research known as transformational leadership. We, thus, approach leadership as both a process of influence intended to change the behavior of followers and as an interactive relational dynamic that involves various members.[13] We draw on several areas of organizational and leadership theory mentioned already, including research on the sources of social power and authority, the nature of the relationship between leader and follower, and the influence process that governs these relationships. But in addition to seeing leadership relationally and as process of influence, we focus on leadership results—the impacts on an organization or institution in ways that make it more effective, cohesive, coherent, or even just. This is to say that we give attention to the role of the institution as part and parcel of the dynamic of leading—a move particularly necessary in academia and academic S&E which have their own context-specific rules and operating principles. So-called "situational" leadership, the idea that different leadership styles are suited to particular contexts and that effective leaders must tailor their style to the situation, grounds our approach to academic leadership for women in S&E.[14]

3.3   LEADING CHANGE

There is one additional aspect of leading that we must mention as it takes into consideration—not only evolving theories—but the needs of women faculty in relation to the changing nature of the college/university setting. James MacGregor Burns in *Leadership* (1978) noted two distinct styles of leading relevant for the mission-based setting of the college or university: transactional and, as mentioned, transformational styles. The transactional leader is the more typical "collegial sort," James Downey notes, one "who nudges, pokes, persuades, occasionally inspires, but generally *manages* his or her institution towards its self-formulated goals and ambitions."[15] The transactional academic

---

[12] D.L. Duke, "The Normative Context of Organizational Leadership," *Educational Administration Quarterly* 34.2 (1998) 165–195.

[13] Rost 1993; James MacGregor Burns, *Leadership* (NY: Harper & Row Publishers, 1978): 18; K. Murrell, "Emergent Theories of Leadership for the Next Century: Towards Relational Concepts," *Organization Development Journal* 15.3 (1997): 35–42.

[14] P. Hersey, K. Blanchard, and D. Johnson, *Management of Organizational Behavior: Leading Human Resources* (Upper Saddle River, NJ: Pearson Education, 2008); F.E. Fiedler, M. Chemers and L. Mahar, *Improving Leadership Effectiveness: The Leader Match Concept* (NY: Wiley, 1978).

[15] James Downey, "Guest Editor's Introduction: Academic Leadership and Organizational Change, *Innovative Higher Education* 25.4 (Summer 2001): 235–238, 236.

leader, in this case, would preserve the status quo, and conduct academic business as usual. By contrast, the transformer, more "clearly" a leader in the visionary sense, is the person "who by dint of talent, character, and charisma give[s] visionary expression and direction to institutional goals and ambitions" (Downey 2001, 236).

In contrast to earlier models, the academy now and over the last decade, Downey observes, has "wish[ed] for transformers," a desire evident in the popularity of Fisher and Koch's *Presidential Leadership: Making a Difference* (1996) which argues for embracing "the bold and skilful exercise of power" to make possible "great advances for the institution" (Downey 2001, 235).[16] This view of the transformational leader "corresponds to the prevailing political and corporate idea of the president as commander-in-chief on whose genius or integrity, or both, rests the fate of the university, the company, or the republic, as the case may be" (Downey 2001, 236). Transformational leadership is, thus, the critical model for "leading change" across the various leadership sectors (whether academia or business management), under the "new rules" of a 21st century leadership climate (i.e., globalization, revolution in information and communication technologies), and under circumstances where change and instability now prevail.

John Kotter describes this philosophy in his analysis of "post-corporate" or change-oriented leadership (much like Rost's postindustrial leader)—a model which is particularly useful for women in academic S&E given his emphasis on nonlinear career paths which women often take.[17] Kotter argues that the "paths to success at work" have radically changed, that "a powerful set of economic forces" have altered the nature of leadership, managerial work, the structure and functioning of organizations, and the increased pace of change—all of which helps those leaders "who can flexibly leverage opportunities," adopt "less linear, less stable career paths," and who are "constantly willing to grow and learn" (1995, 1, 5 & 6). If an ability to "lead change" or transform organizations, as Kotter argues in *What Leaders Really Do* (1999), defines leadership today, women academic leaders in S&E are well-positioned to undertake these roles and in advocating for and tackling institutional change.

Lastly, a current revision of the "typology of transactional and transforming styles" of academic leadership provides women leaders in S&E with a useful conceptual tool for navigating today's complexities of the new university. Both styles, according to Downey, have focused "too much attention on the leader" and thereby "encourage[d] the erroneous belief that organizations rely on a gifted indi-

[16] Downey (2001, 236) notes that the transactional guidebook, Donald Walker's *The Effective Administrator: A Practical Approach to Problem Solving, Decision Making*, and Campus Leadership (1979), which theorizes the college president as chief administrator, went through six printings between 1979 and 1989 but is now no longer available.

[17] John Kotter, *Leading Change* (Boston: Harvard Business School Press, 1996); *The New Rules: How to Succeed in Today's Post-Corporate World* (NY: Free Press, 1995).

vidual" for "their prosperity or even survival"—a premise that "bespeaks a culture of dependence and conformity" at "odds" with how universities actually operate today in "the ideal of highly distributed leadership" (Downey 2001, 237). If universities are to "navigate successfully the straits and shoals posed by today's circumstances," including "the rise of an information economy, changing demographics, new technologies, privatization of higher education, and a convergence of knowledge producing organizations," it will be "by fostering leadership" exercised "at the most strategic points of engagement with challenge" (Downey 2001, 237). These strategic engagements, he continues, will be conducted collaboratively by those "who best understand the forces at work—students, staff, faculty, department chairs, deans, as well as presidents" (Downey 2001, 237). It is in this sense that the visionary leader's role is to take the lead in "cultivating an institutional climate where openness, mutual respect, and the release of creative energies are valued as acts of leadership in themselves" (Downey 2001, 237).

Such leadership priorities—strategic engagement with often systemic challenges, the need for a situational or grounded vision, the importance of coalitions, especially among interdisciplinary and diverse groups, sensitivity to and capacity to lead transformations in institutional climate, and working at points of critical engagement among groups and communities to inspire creative and collaborative innovation—describe the conditions and criteria by which women in academic S&E are becoming leaders today.

3.4   WOMEN AND LEADERSHIP

We now turn to academic leadership with respect to gender and to the significant changes occurring in leadership theories about women's contribution to new models of leadership style and women's advancement and promotion, including in the S&E fields. Though our focus is on academic leadership, since there is not an extensive literature on this area for this particular group, we provide a broader gaze on research in women and leadership to provide an analytical framework for women leaders entering this field.

While by no means a new idea, until fairly recently women leaders were either treated as a contradiction in terms or as a marginalized addition to the field of leadership studies. In this respect, women's status in the literature echoed a material one: while women in the United States make up 46% of managers and administrators (U.S. Bureau of Labor Statistics 2002), for instance, women in top management positions, as The Federal Glass Ceiling Commission (1995) and the Global Fortune 500 (2002) reveal, have made little progress in breaking into the upper-echelons of corporate America. They account for less than 1% of Fortune 500 CEOs, for instance.[18] Political

---

[18] Bureau of Labor Statistics, U.S. Department of Labor, 2002, Household Data: Annual Averages, Table 11: Employed Persons By Detailed Occupation, Sex, Race, And Hispanic Origin, www.bls.gov/cps/cpsaat11.pdf; U.S. Glass Ceiling Commission, *A Solid Investment: Making Full Use of the Nation's Human Capital* (Final Report of the Commission, Washington, D.C., U.S. Government Printing Office, 1995), http://digitalcommons.ilr.cornell.edu/key_workplace/120; The 2002 Global 500: The CEOs. *Fortune*, 2002.

leadership, too, while increasing among women in many nations (Adler 1999), includes few women in the most elite, high-level, or powerful political roles.[19]

In part, women's peripheral place has to do with the sectors traditionally prized in the study and practice of leadership: business leadership, designed to improve corporate efficiency, and political or public sector leadership, intended to promote good governance and policy. More recently, interest in mission-based and nonprofit leadership (including academe, the NGO enterprise, etc.) has increased attention to the role of women—a line of inquiry framed by issues of diversity in leadership style and questions of leadership as a tool of social amelioration.[20] In fact, the study of leadership from the perspective of this subfield—education and academe, non-profits, and NGOs—has come into prominence in part because these areas require distinct models of leadership, offer substantive contributions to leadership studies in general, and, most interestingly, undo the public–private polarization of traditional leadership studies. Indeed, nonprofit leadership has offered overarching and bridge-building principles of good leadership (i.e., an emphasis on communication, values or mission-based leadership, and engaged relationships between leaders and followers), which are useful for a variety of different/non-traditional organizations.[21]

A serious interest in women in this context is evident on several topical fronts: (a.) gender and leadership/managerial style, including the value of "feminized" leadership styles borrowed from studies of corporate restructuring and institutional change; (b.) women, access, and diversity of representation in leadership and organizational culture; and (c.) the role of the nontraditional leader in leveraging transformational or situational leadership skills to foment institutional change. How women academic leaders and administrators fare in traditionally male-dominated fields, including

---

[19] N.J. Adler, "Global Leaders: Women of Influence," in G.N. Powell, ed., *Handbook of Gender and Work* (Sage, Thousand Oaks, CA, 1999) 239–261; Center for the American Woman and Politics, *Women in Elected Office 2008*, Rutgers-The State University of New Jersey, http://www.rci.rutgers.edu/~cawp/Facts.html#elective; United Nations, *The World's Women 2005: Progress in Statistics* (ST/ESA/STAT/SER.K/17) Demographic and Social Statistics Branch, New York, NY, http://unstats.un.org/unsd/demographic/products/indwm/ww2005/tab6.htm; D.T. Miller, B. Taylor, and M.L. Buck, "Gender Gaps: Who Needs to Be Explained? *Journal of Personality and Social Psychology*, 61 (1991) 5–12.

[20] Marlene Fine, "Women, Collaboration, and Social Change: An Ethics-Based Model of Leadership," in eds., Jean Lau Chin, Bernice Lott, Joy Rice, Janice Sanches-Hucles, *Women and Leadership: Transforming Visions and Diverse Voices* (London: Blackwell Publishing, 2007): 177–191; H.S. Astin and C. Leland, *Women of Influence, Women of Vision: A Cross Generational Study of Leaders and Social Change* (San Francisco: Jossey-Bass, 1991); R. Eisler, *The Chalice and the Blade* (Harper San Francisco. 1987).

[21] R.M. Cyert, "Defining Leadership and Explicating the Process," *Nonprofit Management and Leadership*, 1.1 (1990) 29–38; R.P. Chait, W.P. Ryan, and B.E. Taylor, *Governance as Leadership: Reframing the Work of Nonprofit Boards* (NY: John Wiley & Sons, 2005); Burt Nanus and Stephen M. Dobbs, *Leaders Who Make a Difference: Essential Strategies for Meeting the Nonprofit Challenge* (San Francisco: Jossey-Bass, 1999); J. Kouzes and B. Posner, *The Leadership Challenge* (3rd ed., San Francisco: Jossey-Bass, 2003).

the less-often studied fields of science, engineering and technology, is directly relevant to this inquiry.[22]

3.4.1   The Female Advantage? Gender, Style, and Effectiveness

Many studies have explored leadership style and effectiveness in ways attentive to women leaders and to specific positions.[23] For instance, Vicki Rosser et al. (2003) assessed the leadership effectiveness of university deans and directors by relating their leadership behaviors to their perceived effectiveness by those that they supervise. In the university that Rosser et al. (2003) studied, which had an atypically high percentage of women and ethnically diverse administrators, women leaders were rated as more effective than male administrators. One of the underlying questions that this research and similar studies has raised is whether effective academic leadership is gender-based or gender-neutral? It could even be argued that the unique challenges of academic leadership favor the collaborative, consensus-building leadership styles often associated with women's gender socialization.[24]

A number of studies have explored these and additional traits of effectiveness for women in higher education. Astin and Leland's (1991) rich qualitative study of women as "change agents" in higher education, for instance, concluded that collaborative, empowering, envisioning, and networking leadership styles contributed to their effectiveness. Other studies have tapped into broader inquiry about women's traits and capacity as leaders. Women leaders, as compared to men, for instance, have been found to be less corrupt and corruptible (Swamy et al. 2001), less selfish (Eckel and Grossman 1998), and more socially inclined (Eagly and Johnson 1990), including having better social skills.[25]

In the broader literature, additional studies have identified "feminine" or women–specific leadership traits and styles, including what Helgesen (1990) terms "women's ways of leading"—

---

[22] S.V. Rosser, "Senior Women Scientists: Overlooked and Understudied?" *Journal of Women and Minorities in Science and Engineering* 12(4) (2006): 275–293.

[23] W.H. Knight and M.C. Holen, "Leadership and the Perceived Effectiveness of Department Chairpersons," *Journal of Higher Education*, 56.6 (1985) 677–690; V.J. Rosser, L.K. Johnsrud, and R.H. Heck, "Academic Deans and Directors: Assessing their Effectiveness from Individual and Institutional Perspectives," *The Journal of Higher Education* 74.1 (January/February 2003) 1–25; R.B. Womack, "Measuring the Leadership Styles and Scholarly Productivity of Nursing Department Chairpersons," *Journal of Professional Nursing* 12.3 (1996) 133–140.

[24] See Alice H. Eagly, Mary C. Johannesen-Schmidt, and Marloes L. van Engen "Transformational, Transactional, and Laissez-Faire Leadership Styles: A Meta-Analysis Comparing Women and Men," *Psychological Bulletin* 129.4 (2003) 569–591.

[25] Anand Swamy, Stephen Knack, Young Lee, and Omar Azfar, "Gender and Corruption," *Journal of Development Economics* 64 (2001, Feb) 25–55; Catherine Eckel and Philip Grossman, "Are Women Less Selfish Than Men? Evidence from Dictator Experiments," *Economic Journal* 108.448 (1998) 726–735; A.H. Eagly and B.T. Johnson, "Gender and Leadership Style: A Meta-analysis," *Psychological Bulletin*, 108.2 (1990) 233–256.

less-hierarchical chain of command structures, team-building, fostering inclusive networks and relationships, and centering communication and process rather than tasks and outcomes (Claes 1999; Rosener 1990; 1995).[26] Masculine leadership traits, by contrast, include rigid and hierarchical notions of power, centralized, even autocratic authority, and top-down decision-making (Kanter 1977; Eagly and Johannesen-Schmidt 2001; Rosener 1990; Kelly et al. 1991). Some of these very traits, Kellerman has identified as "bad leadership" (2004, 75).[27] While we approach "feminine" and "masculine" as socially constructed attributes, dependent upon culture, context, and conditions, Claes (1999, 436) points out that management theory now generally recognizes these and other cultural "differences" in management styles, i.e., American management style is seen as mostly masculine.

In addition to inquiry into women's capacities as leaders, there are a range of studies that advocate feminized leadership styles or traits for redefining "good leadership" for today's culture and climate. Eagly and Johannesen-Schmidt (2003, 569) observe, for instance, that "[c]laims about the distinctive leadership styles of women abound" today, especially in trade books (e.g., Book 2000; Helgesen 1990; Rosener 1995) and in analyses "that draw on personal experience in organizations and on informal surveys and interviews of managers."[28] A *Business Week* (2000) report from a management research group, for instance, accidentally "stumbled" upon gender differences "while compiling and analyzing performance evaluations." It found that women scored higher in such skill areas as motivating others, fostering communication, producing high quality work, and listening to others, while men tied with women on strategic planning and analyzing issues. A weblink to this report takes readers to a "related" article, "Tips for Male Managers from the Female Playbook," which includes such topics as "controlling bosses are bad bosses" and "so are micro-managers," and such propositions as emphasize "flexibility, not rigidity," "be open to others' opinions," "build consensus," "admit you don't know everything," and "avoid hogging air time—learn to listen."[29]

In general, such sources find women leaders to be "less hierarchical, more cooperative and collaborative, and more oriented to enhancing others' self-worth" than men leaders—patterns of

---

[26] S. Helgesen, *The Female Advantage: Women's Ways of Leadership* (NY: Doubleday Currency, 1990); J.B. Rosener, "Ways Women Lead," *Harvard Business Review*, 68.6 (1990) 119–125; J.B. Rosener, *America's Competitive Secret: Utilizing Women as Management Strategy* (NY: Oxford University Press, 1995); Marie-Therese Claes, "Women, Men and Management Styles," *International Labour Review*, 138.4 (1999) 41–46.

[27] R.M. Kanter, *Men and Women of the Corporation* (NY: Basic Books, 1977); A.H. Eagly and M.C. Johannesen-Schmidt, "The Leadership Styles of Women and Men," *Journal of Social Issues* 57.4 (2001) 781–797; R.M. Kelly, M.M., Hale, and J. Burgess, "Gender and Managerial/Leadership Styles: A Comparison of Arizona Public Administrators," *Women and Politics*, 11 (1991) 19–39.

[28] E.W. Book, *Why The Best Man For The Job Is A Woman* (NY: Harper Collins, 2000).

[29] Rochelle Sharpe, "As Leaders, Women Rule: New Studies Find That Female Managers Outshine Their Male Counterparts in Almost Every Measure," *Business Week*, 20 Nov 2000, www.businessweek.com/2000/00_47/b3708145.htm.

behavior that some argue make "women superior leaders for contemporary organizations" (Eagly and Johannesen-Schmidt 2003, 569).

Likewise, beyond communication and interpersonal skills, Daniel Goleman (1998) finds a gender link to "emotional competency" or emotional intelligence, defined as that ability to be critical self-aware of one's own strengths and limitations as a leader and to the factors and situations that may evoke emotion. These competencies allow leaders to "better manage" emotions and behaviors in relating to individuals and systems.[30] In assessing the leadership competencies of 358 managers across the Johnson & Johnson Consumer & Personal Care Group to distinguish high from average performers, Kathleen Cavallo (2006) found "a strong relationship between superior performing leaders and emotional competence." Her conclusions supported the theory that "the social, emotional and relational competency set commonly referred to as Emotional Intelligence (i.e., self-awareness, self-management, social awareness, and social skills, rated by supervisors and subordinates) is a distinguishing factor in leadership performance."[31] Feminine/feminized leadership traits are now even called for in new corporate environments (i.e., teams, trust-building, social aptitude, interpersonal or emotional intelligence), as traditionally masculine styles are seen as no longer appropriate (Eisler 1997; Goleman 1998).[32]

Discussions of so-called "female advantage," "female excellence in leadership," or the advantage of feminized leadership styles, has even entered mainstream public culture as an emergent sensibility that is gaining ground (Hefferman 2002; Sharpe 2000; Kristof 2008).[33] A recent *New York Times* op-ed article by Nicholas Kristof "When Women Rule," for instance, noted that, "[w]hile no woman has been president of the United States—yet—the world does have several thousand years' worth of experience with female leaders" in which "[t]heir historical record puts men's to shame." Kristof references work in political psychology that finds women excelling in "skills useful in leadership," such as consensus-building, but also laments that such talents, insofar as they contrast with socio-cultural norms and expectations about women leaders, results in a "conundrum" described more than thirty years ago by Rosabeth Moss Kanter: woman leaders face more scrutiny than men and "can be perceived as competent or as likable, but not both" (Kristof 2007).

---

[30] Daniel Goleman, *Working with Emotional Intelligence* (NY: Bantam Books, 1998).

[31] Kathleen Cavallo, "Emotional Competence and Leadership Excellence at Johnson & Johnson: The Emotional Intelligence and Leadership Study," *Europe's Journal of Psychology*, 11 Feb 2006.

[32] Raine Eisler, "The Hidden Subtext for Sustainable Change," in Willis Harman and Maya Porter eds., *The New Business of Business: Sharing Responsibility for a Positive Global Future* (San Francisco, CA, Berrett-Koeler/World Business Academy, 1997).

[33] M. Hefferman, "The Female CEO ca. 2002," *Fast Company*, 61 (Aug 2000) 58–66; N. Zane, "Gender and Leadership: The Need for 'Public Talk' in Building an Organizational Change Agenda," *Diversity Factor* 7.3 (1999) 16–21; Kristof, "When Women Rule," *New York Times*, 8 Feb 2008. See also R. Inglehart and P. Norris (2003), *Rising Tide: Gender Equality and Cultural Change* (NY: Cambridge University Press).

The solution to this conundrum, he continues, is described by economist Esther Duflo's et al.'s (2004) recent work on prejudice, where women leaders chosen by quota systems in village councils in India were judged more negatively by local villagers, even though they performed better by several objective standards. These opinions, born of simple prejudice, Duflo et al. (2004) argue, evident in villagers' habit of rating identical speeches lower when they were delivered by women, were eventually "overridden" with exposure to women leaders and as women leaders were experienced "in action."[34] For Kristof this research offers a lesson about today's culture, that women leaders in "democracies in the television age" face the steep challenge of having to "navigate public prejudices," before enough exposure occurs to allow followers to recalibrate their opinions.

3.4.2   Prejudice, Performance, and Assumptions

While such trends in culture and research are encouraging, the obstacles that women face across various leadership situations, including outright prejudice, are often entrenched, subtle, and serious. Longstanding political research, for instance, indicates that voters not only find women to be worse leaders (Arrow 1973; Phelps 1972) and express unreasoned "distaste" for women leaders (Becker 1957), but that insofar as women leaders are at odds with socio-cultural norms, their presence even reduces the value of this traditionally male activity (Akerlof and Kranton 2000; Eagly and Karau 2002).[35] Additional evidence shows that management ability is linked with "possessing predominantly masculine characteristics" (Powell, Butterfield and Parent 2002, 188; Schein 2001), so that followers often prefer male to female bosses—though this preference is changing somewhat (Powell, Butterfield and Parent 2002; Koch et al. 2005).[36]

---

[34] Raghabendra Chattopadhyay, Esther Duflo, "Women as Policy Makers: Evidence from a Randomized Policy Experiment in India," *Econometrica* 72.5(2004) 1409–1443; Lori Beaman, Raghabendra Chattopadhyay, Esther Duflo, Rohini Pande, and Petia Topalova "Powerful Women: Does Exposure Reduce Prejudice?" http://econ-www .mit.edu/files/2406.

[35] K. Arrow, "The Theory of Discrimination," in O. Ashenfelter and A. Rees eds., *Discrimination in Labor Markets* (Princeton, NJ: Princeton University Press, 1973) 3-33; E. Phelps, "The Statistical Theory of Racism and Sexism," *American Economic Review 62* (1972) 659-661; G. Becker, *The Economics of Discrimination* (Chicago: Chicago University Press, 1957); G. Akerlof and R. Kranton, "Identity and Economics," *The Quarterly Journal of Economics 115*.3 (2000) 715–754; A. Eagly and S. Karau, "Role Congruity Theory of Prejudice toward Female Leaders," *Psychological Review 109* (2002) 573–598.

[36] G.N. Powell, D.A. Butterfield, and J.D. Parent, "Gender and Managerial Stereotypes: Have the Times Changed? *Journal of Management*, 28 (2002) 177–193: 188; V.E. Schein, "A Global Look at Psychological Barriers to Women's Progress in Management, *Journal of Social Issues*, 57 (2001) 675–688; V.E. Schein, "The Relationship between Sex Role Stereotypes and Requisite Management Characteristics," *Journal of Applied Psychology 57* (1973) 95–100.

The implicit masculine image of the leader impacts, not only the perception of women leaders, but women's self-perception (Heilman 2001; Olsson 2000; Nyquist and Spence 1986).[37] As mentioned, Kanter (1977) famously discovered women's lack of aggressiveness and authority, their need to be "nice," and their habit of attributing success to others. Lipsey et al. (1990, 394) have likewise documented what they called "sex role socialization," and Claes (1999, 393) described these and other elements of gender socialization as a "culture trap" for women, resulting in low expectations, fear of success, lack of desire for power, or choosing a dependent role, etc.[38]

More perniciously, if, as Eagly and Karau (2002, 576) note, there is an "incongruity" between "expectations about women," given traditional female gender roles, and about leaders, which "underlie[s] prejudice against female leaders" (see also Burgess and Borgida 1999; Heilman 2001), such beliefs about women and men are activated early on, easily, and often unconsciously by "gender-related cues" that "influence people to perceive individual women" through traditional gender norms. Women, for instance, are viewed as "more communal, selfless," etc., and men are seen as "more agentic" (i.e., assertiveness, instrumentality) (Deaux and Kite 1993; Kunda and Spencer 2003).[39] Because of these and other assumptions about women and their leadership ability, women

---

[37] M.E. Heilman, "Description and Prescription: How Gender Stereotypes Prevent Women's Ascent up the Organizational Ladder," *Journal of Social Issues*, 57 (2001) 657–674; S. Olsson, "Acknowledging the Female Archetype: Women Managers Narratives of Gender," *Women in Management Review* 15.5/6 (2000) 296–302; L.V. Nyquist and J.T. Spence, "Effects of Dispositional Dominance and Sex Role Expectations on Leadership Behaviors," *Journal of Personality and Social Psychology* 50 (1986) 87-93. For a treatment of "stereotype threat," the belief that one's behavior will confirm existing stereotypes which can impact performance, see J. Aronson and C.M. Steele, "Stereotypes and The Fragility of Human Competence, Motivation, and Self-Concept," in C. Dweck and E. Elliot, eds., *Handbook of Competence and Motivation* (NY: Guilford, 2005). But Sabine C. Koch, Rebecca Luft, and Lenelis Kruse, "Women and Leadership—20 Years Later: A Semantic Connotation Study," *Social Science Information* 44.1, 9-39 (2005) 13, 35, found in their semantic study of the perception of leadership, including the term itself, that Kruse and Wintermantel's (1986) original conclusions "that the existence of gender stereotypes affects the perception, description and evaluation of women leaders" has softened so that "leadership ceases to be male and has become more androgynous." L. Kruse and M. Wintermantel, "Leadership Ms.-Qualified: The Gender Bias in Everyday and Scientific Thinking," in C.F. Graumann and S. Moscovici, eds. *Changing Conceptions of Leadership* (New York: Springer-Verlag, 1986): 171–197; E. E. Duehr and J.E. Bono, "Men, Women, and Managers: Are Stereotypes Finally Changing?" *Personnel Psychology* 59.4 (2006) 815–846. For the opposite outcome, see S. Sczesny, "The Perception of Leadership Competence by Female and Male Leaders," *Zeitscrift fur Socialpsychologie* 34 (2003) 133–145; and among young men, D. Jackson, E. Engstrom, T. Emmers-Sommer, "Think Leader, Think Male and Female: Sex vs. Seating Arrangement as Leadership Cues," *Sex Roles* 57.9-10 (2007) 713–723.

[38] Richard Lipsey, Peter Steiner, Douglas Purvis, Paul Courant, *Economics* (NY: Harper & Row, 1990) 394.

[39] Alice H. Eagly and Steven J. Karau, "Role Congruity Theory of Prejudice Toward Female Leaders," *Psychological Review*, 109.3 (2002) 573–598; D. Burgess and E. Borgida, "Who Women Are, Who Women Should Be: Descriptive and Prescriptive Gender Stereotyping In Sex Discrimination," *Psychology, Public Policy, and Law*, 5 (1999) 665–692; K. Deaux and L.L. Lewis, "Components of Gender Stereotypes," *Psychological Documents*, 13.25 (1983) Ms.

are often held to higher standards of performance than men, or they are forced to show superior performance to be viewed as competent (Biernat and Kobrynowicz 1997; Foschi 1996; Shackelford, Wood and Worchel 1996; Wood and Karten 1986).[40] Indeed, confidence and power-seeking behavior when exhibited by women, as Kathleen Hall Jamieson (1997) and Nancy Nichols (1993) note, is perceived in negative terms, given sex/gender cultural norms—even as aggression—thus presenting women leaders with "double-binds" that prevent their managerial effectiveness.[41]

It is perhaps ironic, then, given such perceptions, that most studies also find little actual differences between men and women in leadership performance, or that if women and men do differ in leadership style, they do not really differ in leadership effectiveness (Thompson 2000; Bartol and Martin 1986; van Engen, van der Leeden, and Willemsen 2001).[42] Davidson Cooper (1992), for instance, found that many women managers, when interviewed, possessed the same traits as male managers, while other studies found that female and male managers actually behave in similar ways in leadership roles (Oyster 1992; Gibson 1995).[43]

One important exception is Eagly, Johannesen-Schmidt, and van Engen's (2003, 571) meta-analysis of 45 studies that compared male and female managers on measures of transformational,

---

No. 2583; Crystal L. Hoyt, "The Role of Leadership Efficacy and Stereotype Activation in Women's Identification with Leadership," *Journal of Leadership & Organizational Studies* 11.4 (2005) 2–14; S.J. Spencer, C.M. Steele, D. M. Quinn, "Stereotype Threat and Women's Math Performance, *Journal of Experimental Social Psychology*, 35 (1999) 4–28; Z. Kunda and S.J. Spencer, "When Do Stereotypes Come to Mind and When Do They Color Judgment? A Goal-based Theory of Stereotype Activation and Application," *Psychological Bulletin* 129 (2003) 522–544.

[40] M. Biernat and D. Kobrynowicz, "Gender- and Race-Based Standards of Competence: Lower Minimum Standards But Higher Ability Standards for Devalued Groups," *Journal of Personality and Social Psychology* 72 (1997) 544–557; M. Foschi, "Double Standards in The Evaluation of Men and Women," *Social Psychology Quarterly*, 59 (1996) 237–254; S. Shackelford, W. Wood, and S. Worchel, "Behavioral Styles and the Influence of Women in Mixed-Sex Groups," *Social Psychology Quarterly* 59 (1996) 284–293; W. Wood and S.J. Karten, "Sex Differences in Interaction Style as a Product of Perceived Sex Differences in Competence," *Journal of Personality and Social Psychology* 50 (1996) 341–347.

[41] Kathleen Hall Jamieson, *Beyond the Double Bind: Women and Leadership* (Oxford University Press, 1997); Nancy Nichols, "Whatever Happened to Rosie the Riveter?" *Harvard Business Review*, 71.4 (Jul–Aug 1993) 54–57.

[42] M.D. Thompson, "Gender, Leadership Orientation, and Effectiveness: Testing the Theoretical Model of Bolman & Deal And Quinn," *Sex Roles*, 42.11/12 (2000) 969-992; K.M. Bartol and D.C. Martin, "Women and Men In Task Groups," in R. D. Ashmore and E K. Del Boca, eds., *The Social Psychology of Female-Male Relation: A Critical Analyses of Central Concepts* (Orlando, FL: Academic Press 1986) 259–310; M.L. van Engen, R. van der Leeden, and T. Willemsen, "Gender, Context and Leadership Styles: A Field Study," *Journal of Occupational and Organizational Psychology* 74 (2001) 581–598.

[43] M. Davidson and C.L. Cooper, *Shattering the Glass Ceiling: The Woman Manager* (Paul Chapman, London, 1992); C.K. Oyster, "Perceptions of Power," *Psychology of Women Quarterly* 16 (1992) 527–533; Cristina B. Gibson, "An Investigation of Gender Differences in Leadership across Four Countries," *Journal of International Business Studies* 26 (1995).

transactional, and laissez-faire leadership: they found that women leaders were more transformational, while men typically exhibited more transactional and laissez faire leadership traits in such examples as "attending to followers' mistakes and failures," "waiting for problems to become severe before intervening," or in "exhibiting widespread absence and lack of involvement."[44] The "tendency of women to exceed men on the components of leadership style that relate positively to effectiveness (i.e., transformational leadership and the contingent reward aspect of transactional leadership) and the tendency of men to exceed women on the ineffective styles (i.e., passive management by exception and laissez-faire leadership) attest to women's abilities" and "does suggest female advantage, albeit a small advantage" (Eagly, Johannesen-Schmidt and van Engen 2003, 586)

3.5   STRATEGIC ENGAGEMENTS IN S&E

Women leaders in S&E face additional and, in many respects, unstudied challenges, derived both from working within traditionally masculine or male-dominated organizational contexts and in fields that have historically "gendered" the very nature of the work of science and engineering (Eagly, Makhijani and Klonsky 1992; Madden 2005; Bix 2004).[45] Moreover, the considerable research on gender and leadership, even on women's leadership in male-dominated business and political organizations, is not often applied to women academic leaders in S&E.

Having said this, however, there are sources of information and analytical models useful for approaching this critical need area. Some fruitful lines of inquiry include studies, for instance, that see academia as a crucible for addressing issues of gender, as well as diversity, in leadership, including women's exclusion and inequity (Ottinger and Sikula 1993); research that address the complexity of explaining the underrepresentation of women in academic administrative positions

---

[44] These results are contrary to R.P. Vecchio's claim in "Leadership and Gender Advantage," *The Leadership Quarterly*, 13 (2002) 643–671, that "sex differences in style are an artifact of the placement of women and men in different leadership roles which often cede men more power than women," or are a product of "different standards in judging men and women," Eagly and Carli (2003, 817). See also M. Biernat and D. Kobrynowicz, "Gender- and Race-based Standards of Competence: Lower Minimum Standards but Higher Ability Standards for Devalued Groups," *Journal of Personality and Social Psychology*, 72 (1997) 544–557.

[45] A.H. Eagly, M.G. Makhijani, and B.G. Klonsky, "Gender and the Evaluation of Leaders: A Meta-Analysis," *Psychological Bulletin* 111 (1992), 3–22; Margaret E. Madden, 2004 Division 35 Presidential Address: Gender and Leadership in Higher Education, *Psychology of Women Quarterly* 29.1 (2005), 3–14; A.S. Bix, "From 'Engineeress' to 'Girl Engineers' to 'Good Engineers': A History of Women's U.S. Engineering Education," *NWSA Journal*, 16.1 (2004), 27–49. See also: J.S. Long, ed., *From Scarcity to Visibility: Gender Differences in the Careers of Doctoral Scientists and Engineers* (Washington, DC: National Academy Press, 2001); J.S. Long and M.F. Fox, "Scientific Careers: Universalism and Particularism," *Annual Review of Sociology* 21 (1995):45–71; H. Zuckerman, J. R. Cole and J.T. Bruer, eds. *The Outer Circle: Women in the Scientific Community* (NY: Norton, 1991); G. Sonnert and G. Holton, "The Career Patterns of Men and Women Scientists," *American Scientist* (Jan 1996).

(Neimeier 2004); analysis of women's leadership as an "emergent" leadership style appropriate for higher education (Chliwniak 2000, 2); explorations of the compatibility of new theories of leadership including efficacy, transformational, and implicit leadership with progressive organizational cultures common in academia; investigations of shared or similar leadership challenges in higher education for women and minorities (Aguirre, 2000; Aguirre and Martinez 2002); and critiques of the "masculinized culture" of academia (Swoboda and Vanerbosch 1986) necessitating changes in organizational culture to advance women leaders (Bajdo and Dickson 2001).[46]

Feminist/women's studies theories are helpful and noteworthy in approaching women and diversity in S&E. Beginning with Bleier's (1988, 1) critique of "the so-called objectivity of science and the ways in which science has ignored women, women scientists, and questions which women might ask," Sandra Harding, Ruth Hubbard, Sue V. Rosser, Nancy Tuana, and Anne Fausto-Sterling have argued for "building two-way streets" between the women's studies and scientific communities to rectify the "lack of communication between scientists and feminists"—a gap which can excuse both women's "illiteracy and disinterest" in the sciences and harm "both feminism and science" (Rosser 2002, 66). An ongoing dialogue conducted in the *National Women's Studies Association Journal* demonstrates that "collaboration between women's studies faculty and women scientists can, in fact, help to change the culture of science" (Daly 2004, vii).[47]

But Madden (2005) also notes that this hybrid perspective (women's studies and science, technology, and engineering) has been more fully integrated into building what she calls "a philosophy

---

[46] Cecilia Ottinger and Robin Sikula, "Women in Higher Education: Where Do We Stand?" *Research Briefs* 4.2 (American Council on Education, Washington, D.C., 1993); Luba Chliwniak, "Higher Education Leadership: Analyzing the Gender Gap," *ASHE-ERIC Higher Education Report*, 25.4 (1997); Adalberto Aguirre, "Academic Storytelling: A Critical Race Theory Story of Affirmative Action," *Sociological Perspectives* 43.2 (Summer, 2000): 319–339; A. Aguirre and R. Martinez, "Leadership Practices and Diversity in Higher Education: Transitional and Transformational Frameworks," *Journal of Leadership Studies* 8.3(2002) 53; M. Swoboda and J. Vanderbosch, "The Society of Outsiders: Women in Administration," in P. Farrant ed., *Strategies and Attitudes. Women in Educational Administration* (National Association for Woman Deans, Administrators and Counselors, Washington, D.C., 1986); Linda M. Bajdo and Marcus W. Dickson, "Perceptions of Organizational Culture and Women's Advancement in Organizations: A Cross-Cultural Examination," *Sex Roles* 45.5-6 (Sept 2001) 399–414.

[47] A brief genealogy of the cross-fertilization of women's studies and science, engineering, and technology in the *NWSA Journal* includes: (a.) the most recent Spring 2004 *Special Issue:* (Re)Gendering Science Fields (16.1); (b.) Susan Rosser's 2002 "Twenty-Five Years of NWSA: Have We Built the Two-Way Streets between Women's Studies and Women in Science and Technology?" (14.1); (c.) 2000 *Special Issue:* The Science and Politics of the Search for Sex Differences (12.3); (d.) the 1993 women's studies and science forum by Sandra Harding, Ruth Hubbard, Sue V. Rosser, and Nancy Tuana on "building two-way streets" (5.1); (e.) Ann Fausto-Sterling 's original 1992 article "Building Two-Way Streets: The Case of Feminism and Science" *NWSA Journal* 4.3 (1992):336–349; (f.) Ruth Bleier's 1998 article "The Cultural Price of Social Exclusion: Gender and Science," in the journal's first issue. See also Brenda Daly's "Introduction: *Special Issue:* (Re)Gendering Science Fields," *NWSA Journal* 16.1 (Spring 2004) vi-vii.

of higher education administration," a necessity in the absence of role models, than in transforming S&E proper.[48] In any case, it includes such insights as (a) the influence of sociocultural context on leadership situations; (b) how power dynamics impact sociocultural structures; (c) positing members of the community as active agents of coping and environmental change; (d) the importance of using and valuing multiple perspectives and (e) the necessity of collaboration as an important technique for changing organizations.

One important outcome from this line of inquiry is deeper thinking about the university as an institution with a distinct culture from whose management models, as well as normative ideals, values, and environments, the rest of the management world can learn. In this context, women have proved to be articulate and agile representatives of a higher standard emblematic of mission-based leadership, and educational institutions are seen as offering, not only opportunities for women leaders, but innovative theories and practices of leadership in contrast to the corporate world or public sector.

There is also important work on the unconscious male norm and the masculinized culture of achievement in career paths, especially in academic institutions. Joan Williams notes in *Unbending Gender* (2000, 2), for instance, how organizations often presume the ideal worker as stereotypically male, as "someone who prioritizes work above all other life needs and never takes time off for caregiving," which makes the organization's "systems of advancement and rewards" all based "on the traditional life patterns of men."[49] This creates a "less than welcoming" approach "to the traditional life patterns of women, who have been the primary caregivers in the family" (Williams 2000, 2). Debra Meyerson and Joyce Fletcher (2000, 3) have also developed "the idea that organizations have been created for men on the basis of men's experiences and ideals," so that "both formal and tacit organizational career-advancement systems have been based on the ideal of the married male manager."[50] In this sense, gender *inequity* is "a characteristic of modern organizations" because "a particular view of masculinity shapes the culture and norms" that "prioritizes work above all else, emphasizes individual achievement, and defines success in terms of financial rewards" (2000, 3–4). Using the lives and expectations of white male executives as an implicit norm, organizations then adopt appropriate leadership norms to them i.e., autocratic styles of leadership, prioritizing work over family, aggressive self-promotion (Acker 1998; Williams 2000).[51]

---

[48] See Londa Schiebinger's recent (ed.) *Gendered Innovations in Science and Engineering* (Stanford University Press, 2008) for attempts to move into academic culture change.

[49] Joan Williams, *Unbending Gender: Why Family and Work Conflict and What to Do about it* (Houghton Mifflin Books, 2000).

[50] D. E. Meyerson and J.K. Fletcher, "Modest Manifesto for Shattering the Glass Ceiling," *Harvard Business Review*, 78.1 (2000): 126–136.

[51] Joan Acker, "Hierarchies, Jobs, and Bodies: A Theory of Gendered Organizations," in K. A. Myers, C. D. Anderson, and B. J. Risman eds., *Feminist Foundations: Towards Transforming Sociology* (Thousand Oaks, CA: Sage, 998) 299– 317.

As Marian Ruderman and Patricia Ohlott in *Standing at the Crossroads: Next Steps for High-Achieving Women* note, though diminishing access issues and increasing women in the management pipeline are positive developments, "we still know far more about helping men develop as managers than about helping women in this male-oriented environment."[52] "Conventional career wisdom," they continue, is still based on "the experience of married white men [which] does not readily apply to women," and we still approach women's careers "as exceptions to the male experience" (2002, 4). Instead, they argue, "[w]omen managers and those who work with them need to know how to navigate in the new terrain, and their organizations should know how to develop women in a landscape that is more accepting of them" (2002, 4). In this case, the idea is to open up more gender-neutral understandings of career-advancement in the academic climate—though often the long-term goal of gender-neutrality must be achieved by the short-term focus on gender issues, including barriers to such success.

3.6   WOMEN AS CHANGE AGENTS

In addition to collaborative or relationship-centered forms of leading often associated with women's leadership, the capacity and desire to effect change is often ascribed to women leaders, especially in traditional or male-dominated contexts. While we explore this topic in more detail below, suffice it to say here, as "Advancing Women to Executive Positions" (NAS 2006, 100) notes, that "the presence of a female or a nonwhite in the president's office, "for instance," sends a signal that the campus environment is friendly to women and minorities in a way that brochures and everything else could not send." It is not only that this act "inspire[s] women at all levels of the university, including faculty and students, by demonstrating that women can do as good a job as men," or "they bring unique qualities to the job," but that "as traditional outsiders, women executives may be better able to champion inclusiveness policies and practices" (NAS 2006, 100). In this sense, whether as "traditional outsiders," peripheral or marginalized players in the leadership game, or as activists in institutional climate transformation, women academic leaders in S&E have much to offer.

In the following section, we consider advancing women academic leaders in S&E in more concrete terms, with the insights, programs, and investments in mind that support this endeavor. In addition, we address this nexus of women's representation, participation, and advancement in academic S&E by focusing on women academics as leaders and as potential "change agents."[53]

---

[52] Marian N. Ruderman and Patricia J. Ohlott, *Standing at the Crossroads: Next Steps for High-Achieving Women* (Center for Creative Leadership, 2002): 4.

[53] Myrtle Bell, Mary McLaughlin, and Jennifer Sequeira, "Discrimination, Harassment, and the Glass Ceiling: Women Executives as Change Agents," *Journal of Business Ethics* 37.1 (April 2002): 65-76. For scholars working on empirical demonstrations of this issue, see Nancy Luke and Munshi Kaivan, "Women as Agents of Change: Female Incomes and Household Decisions in South India" (Center for International Development, Harvard University, BREAD Working Paper No. 087, 2004) http://www.cid.harvard.edu/bread/papers/041604_Conference/0504conf/bread_munnar3.pdf.

CHAPTER 4

Women in Engineering Leadership Institute: Critical Issues for Women Academic Engineers as Leaders

4.1   INTRODUCTION: IDENTIFYING CRITICAL ISSUES

This chapter draws from our experiences in developing the leadership program and network for the NSF-sponsored Women in Engineering Leadership Institute (WELI). We identify several critical issues for women academic engineers as leaders in this context. They include:

- The important role of preexisting or simultaneous institutional support for climate transformation, often as a predicate for the success of women leaders in male-dominate fields such as S&E
- Developing a discourse of change based on the experiences of underrepresented groups and using this perspective for reforming attitudes, ideas, and practices among members of this constituency and the university community at large
- Establishing an organization with programs, affiliations, and resources designed to meet the specific needs of women engineers
- Developing a more personalized, informal network of support for members of this group that includes supporters from different communities and positions in both the college/university and beyond (i.e., industry)
- Developing an ideal and practice of transformative leadership that can be shared with the larger community and practiced in an individualized manner by a given leader
- Challenging perspectives that rely on increased representation as the only mark of success, inclusion, and climate change

4.2   DEFINITIONS: MULTIFACETED LEADERSHIP IN ACADEME

In this analysis, we adopt a multifaceted definition of leadership to capture the many different and often nonlinear ways that women lead and come to leadership positions—including the range of experiences that we as a group embody (i.e., program management, department chairs, college deans, directing professional organizations).

By leadership, we include advanced or senior research faculty who often manage teams of students and researchers; department chairs and associate chairs; upper administrators and their associates, including deans, provosts, presidents, and chancellors; program managers and research directors; non-titled or non-positional leaders; and the leadership roles performed in a committee structure and for professional organizations. Each of these positions, despite their situational differences, has in common several core aspects of leading that we have already mentioned: achieving objectives and fostering interactions to benefit an organization (Rost 1993); honing persuasion, influence and other communication skills; practicing vision- and objective-setting; decision making and directing; ensuring the success of an organization and furthering its mission; bringing together coalitions of actors for collaborations (Stogdill 1997); and strategically identifying areas of the institution most primed for or in need of change.

We also use this broad definition of leadership to make an argument for an intrinsic link between excellence and institutional inclusivity, including diversity promotion.[1] There are two aspects to this proposition. On the one hand, we support work that has demonstrated the educational benefits of diversity (Gurin and Hurtado 2002; Gurin and Lopez 2003; Astin 1993). On the other hand, we believe that underrepresented minorities cannot alone be the basis for the broad coalitions needed to transform institutions and, further, that the changes designed to help underrepresented groups invariably improve the environment for all. In this respect, our approach to leadership balances several elements: it is informed by climate, the need for change, and our program development experiences, but it is also designed to improve the total leadership resources available for all members of the university community. By joining interdisciplinary leadership theory with the practical results gleaned from women in S&E leadership programs, including the "lessons learned" from our own NSF-WELI Advanced Leadership & Institutional Transformation Conference, we have found that women academic leaders in S&E have an innovating role to play in the study and practice of leadership.

---

[1] Diversity appears to be an important factor in creativity and productivity. In the corporate world, studies have shown that the companies with more women in leadership positions performed better financially, "The Bottom Line: Connecting Corporate Performance and Gender Diversity," sponsored by BMO Financial Group.

Our first task is to outline—and to synthesize—the insights advanced by important leadership programs for women academic scientists and engineers. We first identify the role and results from the NSF-ADVANCE Programs, including topical themes specific to women academics as leaders in the S&E context. Second, we provide an overview of WELI, emphasizing some of the most critical issues for academic women engineers as leaders. Third, we draw from women and leadership theory to frame these cases to offer an interpretation of "what happened," "what worked," and "why." Lastly, we reflect on how in academic organizational culture women leaders are often faced with multiple institutional challenges, which, in this case, offer insight into the subtle rules of leadership success.

4.3   INSTITUTIONAL TRANSFORMATION: NSF-ADVANCE AND OTHER LEADERSHIP PROGRAMS

One critical catalyst in developing academic leadership programs for underrepresented scientists and engineers has been the NSF's Project ADVANCE: Increasing the Participation and Advancement of Women in Academic Science and Engineering Careers. Begun in 2001, NSF-ADVANCE recognized that the "pursuit of new scientific and engineering knowledge and its use in service to society" requires "talent, perspectives and insight that can only be assured by increasing diversity in the science, engineering, and technological workforce."[2]

This aim was itself gleaned from over 20 years of federally mandated inquiry into the issues, barriers, and challenges for women in S&E and their social implications. This effort also leverages the fact that women now "comprise an increasing percentage of the overall U.S. workforce, and of science and engineering majors at academic institutions," but make up only 27% of the STEM workforce at large, 29% of doctoral S&E faculty at four-year institutions, 18% of S&E full professors at those institutions, with minority women constituting about 3% of S&E faculty at four-year institutions (NSF-ADVANCE, 2). While ADVANCE grants are multiple, including individual career-advancing grants, broad institution-based transformations in infrastructure, and organization-based initiatives, all aim to enable this primary goal of increasing and advancing women and minorities in STEM fields by developing supportive networks, establishing best practices, sharing experiences, building innovative programs and policies, and promoting successful cohorts of women and diverse leaders across institutions and fields. "Creative" and "systemic" solutions are sought from both men and women.

As Table 4.1 outlines, since its inaugural year there have been four "rounds" of ADVANCE proposal requests, with the most recent call (10 August 2007) resulting in awards that have not

[2] NSF-ADVANCE: Increasing the Participation and Advancement of Women in Academic Science and Engineering Careers, NSF 07-582, http://www.nsf.gov/pubs/2007/nsf07582/nsf07582.pdf.

TABLE 4.1: NSF-ADVANCE solicitations					
SOLICITATION	**RP POSTED**	**TYPE**	**FOCUS**	**IT AWARDS**	**EXAMPLE**
Nsf07-582	10 Aug 2007	Round 4	Multidimensional center		Not yet allocated
Nsf05584	15 Apr 2005	Round 3	Single institutional program	13	Cornell University
Nsf02121	8 May 2002	Round 2	Leadership	10	Virginia Tech
Nsf0169	5 Feb 2001	Round 1	Pipeline	9	Georgia Tech

yet been allocated. The ADVANCE awards comprise different categories that have changed over time—including individual leadership awards, institutional transformation awards, programs designed specifically for disseminating new research, and programs for preparing an institution for larger transformation awards. In this analysis, we have focused exclusively on the ADVANCE-Institutional Transformation (IT) award because its objectives are consistent with the emphases here—transforming the climate of an institution to promote the increased participation and advancement of underrepresented academic scientists and engineers, especially through leadership programming.

The ADVANCE-Institutional Transformation (IT) awards are commitments of scale requiring significant research, assessment, and evaluative plans, which means that every awardee is engaged in a systemic self-study of its own institutional climate, including its particular barriers and solutions for greater participation and advancement. It is for this reason that, taken as a whole, the ADVANCE (IT) institutions provide extremely important resources (i.e., online guidebooks for recruiting to expand the applicant pool, manuals for conducting a faculty search that eliminates unconscious bias, gender equity workshops, guides for reviewing tenure and promotion policies and family-work balance, etc.) for assessing the state of research and progress on women and academic leadership in S&E and for aiding others in their attempts to transform their climates.

Some examples of ADVANCE (IT) awards include: Georgia Institute of Technology's ADVANCE Program for Institutional Transformation (Round 1), Virginia Polytechnic Institute and State University's ADVANCE-VT for Institutional Transformation (Round 2), and Cornell University's ACCEL: Advancing Cornell's Commitment to Excellence and Leadership (Round 3).

Each of these programs, for instance, is designed to foster a comprehensive and integrated set of campus-wide initiatives to promote and enhance the careers of women in academic S&E through various levels of institutional intervention—i.e., uncovering systemic barriers, combating unconscious or unintentional bias, revising and reforming university policies and practices, etc. Each program also adopts strategies known to promote change, such as recruitment and retention of outstanding women faculty, establishing supportive networks for mentoring and advising, hosting conferences and workshops, addressing bias in evaluation, promoting gender and leadership seminars, lectures, speaker series, and research initiatives, and collecting data and evaluating programs.

Within the ADVANCE awards, certain ADVANCE-IT programs have developed an intensive focus on leadership training and development. These include the University of Washington-Seattle ADVANCE Center for Institutional Change (CIC), the University of Wisconsin-Madison Women in Science & Engineering Leadership Institute (WISELI), and a smaller leadership-focused initiative at a private institution, Case Western Reserve University's Academic Careers in Engineering & Science (ACES). While there are other programs with leadership dimensions at ADVANCE-IT institutions, these programs have been particularly influential.

The CIC, for instance, is designed to implement program initiatives that eliminate existing barriers to women in science, engineering, and mathematics (SEM) and that "precipitate cultural change at both the departmental and the institutional level." Its focus on "departmental change"—where individual faculty "live or die"—emphasizes that many of the challenges that women faculty face "have their nexus in the department in part because of the strong faculty governance culture in academe." The CIC's two leadership initiatives are a leadership development program for chairs, deans, provost, and the president, and a mentoring program for women faculty in SEM leadership. In addition, the CIC has subsequently created LEAD, a series of national leadership workshops offered annually for department chairs, deans, and emerging leaders from other institutions focused on departmental and university culture and faculty professional development. By contrast, ACES has prioritized a model of institutional transformation in the area of "enhanced transparency in faculty recruitment, advancement, development, and retention policies" and in "improved accountability and effectiveness at the school/college and departmental level." ACES's two leadership program initiates are a pilot-year coaching, mentoring, networking, and training session for women faculty, department chairs, the college dean, and students to bring them "up to speed" regarding the ADVANCE initiative, and an annual Provost Leadership Retreat with current leadership topics of interest to their community.

Despite these different campus priorities, a range of common themes are evident across the ADVANCE-IT leadership initiatives. Table 4.2 summarizes topics from each program's leadership initiatives.

TABLE 4.2: Leadership development workshop topics at WISELI, CIC, ACES

TOPIC (SESSION, PRODUCT, ETC.)	
Leadership training and development • Leadership coaching and development for current deans, chairs, faculty • Things that work, leading change, and creating tomorrow's leaders • Competing for the academic workforce in a global environment • Professional development of senior women faculty; discovery interviews, awards/honors, study of entrepreneurship • Training modules specific to audience i.e., junior, senior faculty, postdocs, engineers, single women, women with children	
Pipeline issues	Building *esprit de corps* in academic departments: How to get to a win–win with faculty and administrators
Climate for women faculty in S&E: • Department cultural change • Improving the chilly climate for women scientists • Discussions of their stories, successes, and suggestions • Affirmative action and organizational change • Tenure-line conversion of non-tenure line women • Harassment and workplace violence	Research on bias and assumptions: • The science glass ceiling • Recognizing and dealing with gender bias • Strategies for countering "the invisible women" • Women in S&E: What the research really says
Cumulating disadvantages: Gender stereotypes, 'small inequalities,' and women's careers	Recruiting for diversity: Deconstructing past recruitment seasons
Getting our voices heard and gendered communication in academe: • Communication between men and women • Voice training • Delivering bad news: Giving negative feedback and having difficult conversations • Patterns of participation in university meetings • Strategic use of humor	Faculty retention strategies • Women faculty mentoring program • Mentoring women for leadership • Developing networks

TABLE 4.2: *(Continued)*	
TOPIC (SESSION, PRODUCT, ETC.)	
▪ How to say "no" and still be seen as a team player ▪ Presenting yourself effectively	
Negotiation	Diversity/sensitivity training
Dual career policy and hires	Research on student ratings of women faculty: Race, gender and personality, or Putting student evaluations of faculty in perspective
Success strategies for women in academic careers: ▪ Sharing success strategies of senior S&E women faculty ▪ Mid-level transitions: associate to full professor ▪ Transitions through the academic career in the tenure and promotion process ▪ How to use national professional meetings to maximize your chance of promotion	Promoting your faculty: ▪ Awards, honors and nominations ▪ Conducting faculty merit reviews ▪ Preparing faculty for promotion to full professor
Department Chair leading strategies ▪ Leadership succession planning ▪ Supervising non-academic staff ▪ Using emotional intelligence ▪ Increasing the visibility of women ▪ Chairing a meeting ▪ The politics of search committees ▪ Negotiating for resources ▪ How to challenge unfair manuscript reviews ▪ How to be an effective manager	Work-life balance, family leave and tenure clock extensions ▪ How to use flexible policy options to support faculty ▪ Work/life balance for faculty: Research and recommendations on family-friendly policies and practices ▪ Faculty work life ▪ Childcare ▪ Science faculty talk about self, home, and career ▪ Time and project management ▪ Balancing a research career and your personal life

These commonalities in programmatic efforts reflect research findings in the literature and from local climate assessments. These include: an emphasis on faculty recruitment, retention and promotion strategies (including mentoring and teaching support); work-life balance; developing leadership and management skills (communication, negotiation, faculty support, etc.); critically evaluating department and institution-wide policies, practices, and procedures, including uncovering unconscious bias, the glass ceiling, accumulated disadvantages, and other institutional subtleties; developing and promoting faculty and graduate students including coaching and mentoring and promoting professional self-confidence and self-efficacy; and implementing change, including developing new institutional resources and positions.

Table 4.3 provides examples of additional leadership development programs that work within the same thematic concerns supported by the NSF and/or the ADVANCE Project.

4.4   THE WOMEN IN ENGINEERING LEADERSHIP INSTITUTE

Originally titled the "Women Engineering Faculty Leadership Network," the Women in Engineering Leadership Institute (WELI) also emerged from the ADVANCE initiative. Although it was designed to meet the specific needs of women academic engineers, women scientists in and outside of academia have been important participants from the program's beginning. In certain respects, WELI capitalized on one significant insight resulting from ADVANCE program evaluations: the realization that institutional change occurs not only from "below," from the many participants in a given system, but also from "above," from the work of institutional leaders. Thus, in confronting

TABLE 4.3: Additional leadership development programs	
PROGRAM	LOCATION
COACh (Committee on the Advancement of Women Chemists)	Multi-institutional: Committee on the Advancement of Women Chemists
LEAD	University of Washington-Seattle
PFED	University of California-Irvine
Workshops for Junior Faculty (GEP)	Hunter-CUNY
ALP (Academic Leadership Program) DEO (Department Executive Officers Program)	CIC campus-consortium of 12 research universities, 11 members of the Big Ten Conference and the University of Chicago

the stubborn problem of women and minority underrepresentation in engineering, WELI began with the premise that academic leadership—and training and supporting women academic leaders in engineering—would enable them to play a powerful and important role in institutional change (WEFN 2004).[3]

Established in 2003, the idea for WELI emerged in the aftermath of an earlier NSF-sponsored Leadership Conference in 2000 in Winter Park, Colorado, organized by Deb Niemeier, of the University of California-Davis, and Delcie Durham, Priscilla Nelson, and Alison Flatau of NSF. Valerie Davidson describes the atmosphere of the original conference in this way:

> I was 1 of 32 women chosen out of a pool of over 180 applications to attend the conference... Academic leaders (all women) were invited to make presentations but in a format that emphasized active discussion with all participants. The idea of bringing women together to focus on issues of academic leadership in engineering grew out of discussions among a small group of women. These were women who had moved into senior academic roles and senior positions at NSF. Their concern, based on their own experiences as well as knowledge of the academic culture, was that women were moving into faculty positions but that department chair would be a "glass ceiling" for women. They also recognized that experience as department chair was important for positions in upper administration. So they felt pro-active steps were necessary to effect changes.

Women engineering faculty began to articulate the need for WELI, to build a consensus and "unanimous support" for "a vehicle to coordinate activities conducive to increasing the number of women pursuing academic leadership opportunities in engineering" (WEFN 2004). A proposal for WELI was developed in February 2001, along with the election of an Executive Board of 7 members, including first President Judy Vance of Iowa State University, where WELI's administration was subsequently located. WELI was funded by Project ADVANCE from August 2003 to July 2007. The main initiatives outlined in the proposal included developing and hosting several national Leadership Conferences, Advanced Leadership Workshops, and a Leadership Summit, as well as establishing an online network and community of support for women engineering faculty. In addition, WELI was conceived from the get go as a network, with participation from many different academic institutions and as a space to bridge the gap between academia and industry in the S&E context.

---

[3] NSF-ADVANCE Award Abstract #0245084, Collaborative Research: The Women Engineering Faculty Leadership Network, (11 Aug 2003 to 31 Jul 2007) Iowa State University.

The core premise of WELI was the concern for the relative scarcity and isolation of women engineering faculty at all levels, but especially those in the mid-career and the senior level ranks and in administrative leadership positions. According to Kang (2003), 2.8% of all full professors in engineering are women, and 9.4% of all associate professors in engineering are women. To highlight the meaning of these numbers, Kang (2003) noted that "for every 200 engineering faculty, only seven are women full professors."[4] In addition to low numbers of full professors, women academic engineers were also only very infrequently leading their departments. The following numbers among AAU department chairs in the summer of 2000 (Niemeier and Gonzalez 2001) illustrates this situation comparatively:

FIELD	NUMBER OF DEPARTMENTS	NUMBER OF WOMEN CHAIRS	PERCENTAGE
Engineering	273	6	2.2
Earth Sciences	58	7	12
Chemistry	60	3	5

These conditions—very low numbers, isolation, few opportunities for leadership—among women engineering faculty made "change slow and difficult" (WEFN 2004). These circumstances necessitated taking concrete steps to "connect women engineering faculty members from around the country to discuss and explore the challenges and opportunities of academic leadership" (WEFN 2004).

WELI was established to address these and additional issues of women's participation and advancement in engineering. As the original NSF proposal stated, not only did WELI seek to bring women faculty "together in a supportive environment to learn academic leadership techniques," especially "those skills they may not see in their male-dominated faculty," but to "partner with successful women academic leaders who can help prepare them for the challenges of academic leadership" (WEFN 2004). Indeed, this initiative was seen as urgent for women engineering faculty among the S&E fields since they are "further isolated because they are dispersed within individual engineering departments."

In addition, and this was WELI's most important contribution, WELI distinguished itself from other ADVANCE-IT programs aimed at changing *individual* institutions by advocating, instead, "cross-institutional efforts" for producing "significant individual institutional changes" (WEFN 2004).

---

[4] Kelly Kang, *Characteristics of Doctoral Scientists and Engineers in the United States: 2001*, NSF03-310 (Arlington, VA: National Science Foundation, Division of Science Resources Statistics).

WELI OBJECTIVES AND MISSION

1. Increase the number of women in academic leadership positions
2. Accelerate and enhance the success of women in academic leadership positions
3. Establish and maintain a support network for women engineering faculty leaders
4. Provide support, including training, mentoring, and networking opportunities in academic leadership, for women engineering faculty

4.4.1 WELI's Unique Format

If WELI advocated leadership as a critical mode of institutional change and, further, promoted trans-institutional practice as necessary to make change significant and durable, one of the most important means for achieving these goals was WELI's unique format and approach. WELI described its mission as facilitating "grass roots" institutional change by preparing a larger pool of women engineering faculty to enter the academic senior and administrative ranks. In this respect, WELI was deliberately created to "develop a group of highly capable women engineering faculty and ensure that they are prepared to assume academic leadership roles across the country where they can transform engineering education both through their presence and by taking initiative to increase diversity."[5] To do this, a series of Gordon-style national leadership conferences were developed "to ensure that women faculty would have equal access to the knowledge, preparation, and mentoring required for assuming leadership positions by fostering networking across all types of institutions."[6] The Gordon Research Conferences are a well-known institution among scientists. Designed to promote intense discussion at the frontiers of science, they were developed by Neil E. Gordon at Johns Hopkins University in the late 1920s out of a concern that the traditional means for disseminating information (i.e., publications, presentations at large scientific meetings) were not as effective or dynamic in advancing research. They are also private and confidential: participants agree to treat all discussions as private communications not for public use, including in subsequent publications. Essentially, WELI utilized the conference setting as a vehicle to develop an alternative vision of what academic engineering could and should be.

---

[5] http://www.weli.eng.iastate.edu/about_weli.htm

[6] For a description of the Gordon Research Conference, see http://www.grc.org/about.aspx. In general these conferences aim to "promote discussions and the free exchange of ideas" at the frontiers of research. They involve "scientists with common professional interests come together for a full week of intense discussion and examination of the most advanced aspects of their field" and provide a venue for stimulating and disseminating ideas "that cannot be achieved through the usual channels of communication, i.e., publications and presentations at large scientific meetings." In addition, however, conference participants agree that information presented in all formal and informal sessions is a private communication and restricted from public use.

TABLE 4.4: Summary of the leadership workshops developed and implemented by WELI

DATE	EVENT	LOCATION	PARTICIPANTS
Oct 11–15, 2000	Women in Engineering Leadership Conference	Winter Park, CO	37
Oct 9–11, 2003	Academic Women in Engineering Leadership Institute Workshop/SWE 2003 National Conference	Birmingham, Alabama	21
Nov 5–8, 2003	Leadership Development Conference	University of Utah, Snowbird, Utah	33
May 3–5, 2004	Leadership Summit	University of Connecticut, Storrs, Connecticut	62
Oct 7–10, 2004	Institutional Transformation/ Advanced Leadership Workshop	Syracuse University, Syracuse, New York	32
Apr 28–May 1, 2005	Leadership Development Conference	University of Central Florida, Cocoa Beach, Florida	44
Nov 3–5, 2005	WELI Leadership Workshop with SWE 2005 National Conference	Anaheim, CA	38
Oct 12–14, 2006	WELI Leadership Workshop with SWE 2006 National Conference	Kansas City, MO	
Jun 11–14, 2006	WELI Leadership Workshop with WEPAN 2006 National Conference	Pittsburgh, Pennsylvania	

But WELI also adapted and modified the open communication and intense discussion format of the Gordon conference to the needs of women in academic engineering. The WELI Conferences were designed to foster an informal "workshop" atmosphere where women faculty could 'connect' by unabashedly engaging in discussions about advancement and leadership at all levels: sharing and critically reflecting upon their experiences in various settings (department, university) and in positions of leadership: planning for future career advances, including specific administrative positions; developing leadership skills and a personal leadership style or philosophy; coordinating and collaborating with other women on the various issues raised during the conference; and disseminating these "lessons learned" to other colleagues in other settings, as well as one's home institution.

There were three main groups of activities promoted at each WELI Conference: (a) leadership training and development seminars and lectures; (b) interaction with current women S&E administrators in academia and industry; and (c) formal and informal networking activities. Likewise, the following strategies were consistently used at the WELI Conferences:

- facilitating and stimulating open discussion of the challenges and opportunities of academic leadership
- exploring how transformative leadership methods can be used to usher in a new model of leadership for engineering faculty that would support and strengthen women engineering faculty leaders and expand the community of engineers
- providing the background, knowledge, and support systems needed to effectively compete for and succeed in academic leadership positions
- creating a nationwide network of organizations willing to share insights about their efforts to support women engineers, both in academia and industry

The WELI Leadership Summit was designed to be an aggregate event where leaders from various organizations and participants from past WELI events could discuss action items necessary to increase and advance women in engineering, especially academia. See Table 4.4 for the WELI-sponsored yearly conferences, as well as the WELI-affiliated conferences, and the institutions that hosted them.

The six WELI Leadership Conferences/Workshops (including the Summit)—and not including the 2006 WELI-affiliated Conferences—involved over 200 participants representing over 124 institutions, 39 U.S. states, Washington D.C., Puerto Rico, New Zealand, and Canada. Pre- and post-conference surveys completed by many participants were developed and subsequently analyzed by Deb Niemeier.[7] Most of the assessment questions evaluated the programs and activities

---

[7] D.A. Niemeier (2005a), *Conference Evaluation. Summary of Findings: Pre-Conference and Post-Conference Surveys* (NSF Women in Engineering Leadership Development Conference, Cocoa Beach, FL.: Oct 2005); D.A. Niemeier (2005b), *Conference Evaluation. Summary of Findings: Pre-Conference and Post-Conference Surveys* (NSF Women in Engineering

EVENT	TOTAL	FULL PROF	ASSOC PROF	ASST PROF	OTHER
TABLE 4.5: WELI leadership conference/workshop participants (Niemeier 2004a; 2004b; 2005a; 2005b)					
2003 Alabama Leadership Workshop	21	3	4	0	14
2003 Utah Leadership Conference	30	8	18	4	0
2004 Connecticut Leadership Summit	75	22	16	4	33
2004 New York Advanced Leadership Conference	33	26	5	0	2
2005 Florida Leadership Conference	44	10	34	0	0
2005 California WELI-SWE Workshop	38	13	25	0	0

at a particular conference, each of which had different goals and emphasizes. For the participant makeup of the WELI Conferences by rank, see Table 4.5.

There are, however, strong indicators of the strength of the conference/workshop experience for participants across these events. See Table 4.6 for answers by attendees to the following overarching questions: "Would Encourage Others to Attend Similar Conferences," "Would Support Holding Additional Conferences," "Feel Part of Stronger Network," and "Better Understanding of Academic Leadership." Answers to these questions were rated on a five-point scale, where 1 represented "strong agreement" and 5 represented "strong disagreement" (Niemeier 2004a; 2004b; 2005a; 2005b). In the case of the Utah Conference, for instance, 24 of 30 attendees "Would Encourage Others to Attend Similar Conferences" and 22 "strongly agreed," whereas 24 of 30 "agreed" that they "Would Support Holding Additional Conferences," and 16 "strongly agreed" and 7 "agreed" of 30 that they "Feel Part of Stronger Network."

In addition, in October 2005, WELI's home-office at Iowa State University emailed all past WELI participants with a follow-up assessment questionnaire which returned approximately a 50% response rate.[8] Of these answers, over 70% of the respondents identified "networking with other

---

Leadership Institute: The Advanced Leadership Workshop, Syracuse, NY, Apr 2005); D. A. Niemeier (2004a), *Conference Evaluation, Summary of Findings: Pre-Conference and Post-Conference Surveys* (NSF Women in Engineering Leadership Development Conference, Salt Lake City, Mar 2004); D.A. Niemeier (2004b), *Conference Evaluation, Summary of Findings: Pre-Conference and Post-Conference Surveys* (NSF Women in Engineering Leadership Summit, Storrs, CT, Jun 2004). See WELI website: http://www.weli.eng.iastate.edu/Resources/weli_reports.html.

[8] http://www.weli.eng.iastate.edu/About_WELI/grant_pis.htm

TABLE 4.6: WELI selected conference/workshop assessments (Niemeier 2004a, 2004b, 2005a, 2005b)			
QUESTIONS: "STRONGLY AGREED" AND "AGREED"	CONFERENCE/WORKSHOP: NUMBER OUT OF TOTAL NUMBER		
Would Encourage Others to Attend Similar Conferences	24/30 Utah	23/33 NY	42/44 FL
Would Support Holding Additional Conferences	24/30 Utah	21/33 NY	42/44 FL
Feel Part of Stronger Network	23/30 Utah	20/33 NY	38/44 FL
Better Understanding of Academic Leadership	21/30 Utah	20/33 NY	41/44 FL

women" as "the most important benefit" gleaned from attending a respective WELI Conference, though "learning leadership skills" and "interacting with existing women administrators" were also popular responses to this question. This assessment also identified two persistent challenges consistently identified by women engineering faculty: feeling isolated and unaware of "the opportunities available in leadership positions." Indeed, this common perspective was expressed in this way: the "biggest benefit was the networking aspect, as I have been the only female in my department my entire career." The questionnaire also asked participants to identify "leadership activities" that were "performed as a result of attending a WELI Conference," and, further, whether "the Conference entered into the decision to pursue leadership positions." Many respondents identified these following leadership roles and activities that were pursued post-conference:

- leading ADVANCE grant proposals
- chairing committees
- applying for leadership positions
- becoming department chairs
- chairing diversity committees
- chairing search committees

Most respondents also stated that "WELI had increased their confidence in their ability to take on leadership roles." Respondents were also asked to address the "value" of connecting and networking with others at the conference, and their answers were then categorized into two general responses: participants felt that they were part of a community, that "they knew more people across the country," and they felt that they now shared similar issues with these faculty members. One respondent, for instance, echoed this common sentiment: the "information you receive at a conference is extremely difficult or impossible to get elsewhere," and the "WELI conferences remain the most valuable conferences I have ever attended." Respondents also mentioned that the "Workshops such as these ought to be continued indefinitely," and that "these are important and invigorating events to continue."

4.5   BUILDING AN ALTERNATIVE NETWORK

At the heart of the WELI contribution is the creation of an innovative network—an advanced community to encourage, support, and resource women engineering faculty in the process of advancing their careers in various climates. This key objective emerged in concrete form from both conference and virtual venues. The electronic venues, the WELI webpage and listserv, both managed at Iowa State University, were especially helpful in provided news and links to WELI and women in S&E related events, opportunity postings, data, and research literature. But they also provided a space for ongoing interactions over shared interest areas, including work-life balance strategies, legal and Title IX issues, job announcements, problem solving in administrative positions, etc. As many respondents to the follow-up questionnaire issued by WELI noted, "meeting other ambitious talented women" was one of the biggest benefits of conference attendance.[9]

It is important to briefly revisit the key role of the network as established in research in the multilayered process of career advancement. "The central argument of network research," according to Daniel J. Brass et al. (2005, 795), is that "actors are embedded in networks of interconnected social relationships that offer opportunities for and constraints on behavior."[10] Informal networks are especially important because they "direct the flow of information and power in work organizations," often by "circumvent[ing] formal procedures," and they provide special access to jobs, tools for advancing up the corporate ladder, skills, and even means for acquiring legitimacy.[11] But, networks also function differently for men and underrepresented groups. As much research shows, "women and minorities have limited access to or are excluded from organizational

---

[9] http://www.weli.eng.iastate.edu/About_WELI/grant_pis.htm

[10] Daniel J. Brass, Joseph Galaskiewicz, Henrich R. Greve, and Wenpin Tsai, *Academy of Management Journal* 47.6 (2004) 795–817.

[11] Gail M. McGuire, "Gender, Race, and the Shadow Structure: A Study of Informal Networks and Inequality in a Work Organization," *Gender and Society* 16.3 (Jun 2002) 303–322: 304. See also K.E. Campbell, "Gender Differences in Job-related Networks," *Work and Occupations* 15 (1988): 179–200; K.E. Campbell, P. Marsden, and J. Hurlbert, "Social Resources and Socioeconomic Status," *Social Networks* 8 (1986):97–117; K.E. Campbell, and R.A. Rosenfeld, "Job Search and Job Mobility: Sex and Race Differences," *Research in the Sociology of Work* 3 (1985):147–174; J.M. Podolny and J. N. Baron, "Resources and Relationships: Social Networks and Mobility in the Workplace," *American Sociological Review* 62 (1997):673–693.

networks."[12] Gail McGuire (2002, 304), for instance, emphases that the network represents the "informal side of organizational life," the place "where unspoken rules of interaction" reign and where gender inequality may therefore become both "possible" and "highly resistant to change."[13] This occurs precisely because networks are where a "variety of instrumental resources that are critical for job effectiveness and career advancement," as well as "expressive benefits such as friendship and social support," are allocated (Ibarra 1993, 56).[14] Indeed, as Herminia Ibarra notes, "exclusion or limited access" to informal networks "has long been one of the most frequently reported problems faced by women and racial minorities in organizational settings" (1993, 56).[15] As early as the 1970s, Rosabeth Moss Kanter (1977, 164) had observed in *Men and Women of the Corporation* that "[s]omewhere behind the formal organization chart [of a given firm] was another, a shadow structure in which dramas of power were played out." Thus, limited network access can "produce multiple disadvantages," including "restricted knowledge of what is going on" in an organization, "difficulty forming alliances," both of which are, in turn, associated with "limited mobility and 'glass ceiling' effects" (Ibarra 1993, 56).

If Kanter's notion of the network as a shadow structure in which dramas of power are played out is a particularly evocative image, in many respects WELI was developed to undercut such a shadow structure and to offer a more inclusive alternative network to support women and underrepresented members of the university community. This connected community is increasingly important in leadership development, especially as recent research has suggested that networks positively impact leadership effectiveness.[16] Likewise, Ibarra (1993) found links between networks and the organizational context in which they are embedded, a nexus that because it is one source for the "unique constraints on women and racial minorities," can be the potential means by which women and minorities become active agents (1993, 56). In fact, Laurel Smith-Doerr (2004, 25), who specifically studied women's careers in the life sciences, found that the network could explain why women have more opportunities in certain employment settings: specifically the organizational

---

[12] Herminia Ibarra, "Personal Networks of Women and Minorities in Management: A Conceptual Framework," *The Academy of Management Review* 18.1 (Jan 1993) 56–87.

[13] Gail M. McGuire, "Gender, Race, and the Shadow Structure," 2002: 304.

[14] N.M. Tichy, "Networks in Organizations," in P.C. Nystrom and W.H. Starbuck, eds., *Handbook of Organization Design* (NY: Oxford University Press 1981): 225–248.

[15] Herminia Ibarra, "Personal Networks," 1993: 56. See also: N. DiTomaso, D.E. Thompson, and D.H. Blake, "Corporate Perspectives on the Advancement of Minority Managers," in D.E. Thompson and N. DiTomaso, eds., *Ensuring Minority Success in Corporate Management* (NY: Plenum Press 1988) 119–136; A.M. Morrison and M.A. Von Glinow, "Women and Minorities in Management," *American Psychologist* 45 (1990): 200–208.

[16] Prasad Balkundi and Martin Kilduff, "The Ties that Lead: A Social Network Approach to Leadership," *The Leadership Quarterly* 17 (2006) 419–439.

form of the network in contrast to the hierarchical bureaucracy could alleviate gender stratification and allow for greater equity, especially in advancement and promotion.[17] "Hierarchy and rules hide gender bias," she noted, while "reliance on ties outside the organization provides transparency and flexibility," so that women scientists were "nearly eight times more likely to supervise in the network-oriented biotech firms than in more hierarchical settings" (2004, 25). If networks enable power, performance, and resources in the work context, then this occurs in predominantly three ways: information and resource access; mentorship, advising and guidance; and inclusion into the processes of power, decision-making, and influence.

Ibarra, however, offers several cautions about the prospects of developing alternative networks for women and minorities. The issue is complicated by several factors. On the one hand, while networks are advantageous for underrepresented groups in an organization, research suggests that in both peer (South et al. 1982) and superior-subordinate relationships (Kram 1988; Thomas 1990; Tsui and O'Reilly 1989), "cross-sex/cross-race ties tend to be weaker than homophilous relationships"—that is relationships where the interacting individuals are similar in identity or organizational group affiliations (Ibarra 1993, 68).[18] Apparently "interpersonal similarity increases ease of communication, improves predictability of behavior, and fosters relationships of trust and reciprocity" (Ibarra 1993, 61; Kanter 1977; Lincoln and Miller 1979).[19] If a given underrepresented manager/leader, for instance, is "structurally constrained" to have more cross-sex or cross-race network relationships, perhaps because there are very few people like her in her organization, then her personal networks will also "be characterized by a preponderance of weak, uniplex ties" (Ibarra 1993, 68). In addition, Ibarra (1993, 69) notes that personal networks are "shaped by stereotypes, attributions, and biases that are bolstered by structural arrangements," which "impose significant limits on women's and minorities' abilities to develop the more available and instrumentally usefully heterophilous ties."

If, however, women and minorities then form their own networks, there are certain limitations to them. The most obvious limitation is number. The "demography of a social system," Ibarra

[17] Laurel Smith-Doerr, "Flexibility and Fairness: Effects of the Network Form of Organization on Gender Equity in Life Science Careers," *Sociological Perspectives* 47.1 (2004) 25–54: 25.

[18] S.J. South, C.M. Bonjean, W.T. Markham, and J. Corder, "Social Structure and Intergroup Interaction: Men and Women of the Federal Bureaucracy," *American Sociological Review* 47 (1982): 587–599; K.E. Kram, *Mentoring at Work: Developmental Relationships in Organizational Life* (NY: University Press of America, 1988); D.A. Thomas, "The Impact of Race on Managers' Experience of Developmental Relationships (Mentoring and Sponsorship): An Intra-Organizational Study, *Journal of Organizational Behavior* 2 (1990): 479–492; A.S. Tsui and C.A. O'Reilly, "Beyond Simple Demographic Effects: The Importance of Relational Demography in Superior-Subordinate Dyads, *Academy of Management Journal* 32 (1989): 402–42.

[19] J.R. Lincoln and J. Miller, "Work and Friendship Ties in Organizations: A Comparative Analysis of Relational Networks," *Administrative Science Quarterly* 24 (1979): 181–199.

notes, is a "constraint" on individual preferences for a given network, since women and minorities "usually have a much smaller set of 'similar others' from whom to develop professional relationships based on identity-group homophily"—which means that the personal networks of both male and female entrepreneurs "tend to contain mostly men" (Ibarra 1993, 67; Aldrich, Reese and Dubini 1989).[20] Ibarra also notes that in many organizational contexts, if women and racial minorities "desire network contact with members of their own identity group, they are likely to have to reach out further in their organizations, beyond their immediate peers, superiors, subordinates, or functional areas" (Ibarra 1993, 67; Thomas 1990). Ibarra (1993, 68–9) thus concludes that in "the demography of the average American corporation … homophilous ties are less available, have less instrumental value, and require more time and effort to maintain (due to dispersion and turnover) for women and minorities than for their white male counterparts," while at the same time, "a preponderance of heterophilous contacts, in turn, reduces the likelihood of strong ties as well as the interconnectedness of parties to the network." These are potentially serious obstacles to any alternative network.

In developing women-based networks in S&E, these considerations must be taken into account. Yet, context also figures into the question, since some of the structural constraints that are so powerful in the development and functioning of networks in the corporate context may be significantly different in academia, where more flexible and less hierarchical interaction patterns may occur. Indeed, beyond context, colleges and universities are often invested in fostering "heterophilous" and "cross-sex or cross-race" ties and relationships, which may then alleviate some of the pressures on individuals to choose network interactions and relationships among similar group members. Further, once an individual chooses more cross-sex or cross-race network relationships, her personal networks may not always be characterized "by a preponderance of weak, uniplex ties," since in the academic context, these types of relationships may receive a positive valence (Ibarra 1993, 68). Nevertheless, in the case of developing an alternative network, as WELI has, there are structural and administrative vulnerabilities that this analysis helps us to identify. For instance, there is no doubt that women engineers have a much smaller available pool of colleagues to network with, not only in the case of other women academic engineers, but especially in the category of senior women faculty and administrative leaders. In this respect, WELI has consistently targeted women academic scientists as well as engineers, including those beyond academia. Moreover, the intrinsic limits of potential members also means that the WELI network will and must "contain men," a reality which suggests the value and need for more deliberate thinking about how to include non-similar members into the networking process—both so that the "heterophilous" ties can be strengthened and so that men scientists and engineers can be supporters and advocates of institutional transformation.

[20] H. Aldrich, P.R. Reese, and P. Dubini, "Women on the Verge of a Breakthrough: Networking among Entrepreneurs in the United States and Italy," *Entrepreneurship and Regional Development* 1(1989): 339–356.

Lastly, as Ibarra also suggests, underrepresented groups who "desire network contact with members of their own identity group" will have to reach farther afield, "beyond their immediate peers, superiors, subordinates, or functional areas." In this respect, WELI has been particularly effective in developing the multi-institutional and virtual aspects of its network and, as Table 4.4 indicates, even linked its activities with well-established professional organizations and their national conferences. In this respect, WELI has cultivated many places for disseminating the value of its network as a resource for women academic scientists and engineers.

But perhaps the most critical issue in developing an alternative network is the suggestion that in the typical organizational setting homophilous ties will not only be "less available," but they will have "less instrumental value" and "require more time and effort to maintain" for women and minorities. This insight is in many ways most descriptive of the challenges confronting WELI for the future—the amount of time, effort, and personal initiative required by women academic engineers, often at the expense of their own research and teaching, in maintaining the network. These challenges include finding WELI a sustainable institutional location, sustaining and building membership, cultivating new leadership, and garnering funding and support. In this respect, institutional transformation itself is a prerequisite for the success of this alternative network.

4.6   TRANSFORMATIVE THEORIES OF LEADERSHIP

If WELI offers a distinctive format and approach different from other leadership program efforts, it also provides innovative content, particularly in its theory of leadership. Two precepts of leadership theory—transformative leadership and the role of "instigators"—have informed WELI. While we have discussed the term "transformational" in leadership theory, in the WELI context the idea was used in an additional sense: not only as a model of leadership style, but as an institutional strategy. In this case, the idea of transformative leadership was drawn from the institutional transformation associated with the highly successful Kellogg National Fellowship Program (1980–2001), which attempted to develop individuals as leaders to produce "transforming changes" in communities, agencies, and institutions at all levels of governance (Markus 2001).[21] This use of the term is also evident in ADVANCE's notion of the role of the leader in helping to conduct climate change in academic policies and procedures, as well as in increasing the ranks of women science and engineering faculty. Lastly, the institutional sense of transformative leadership refers to a notion of sustainable or lasting change that requires moving beyond traditional organizational theory to consider broader, multilateral influences, including those that arise from social movements

[21] Gregory Markus, "Building Leadership: Findings from a Longitudinal Evaluation of the Kellogg National Fellowship Program," (W.K. Kellogg Foundation, Battle Creek, 2001).

motivated by marginalized groups or visionary leaders (Olsen 1965; Clemens 1993; MIT 1999).[22] On this last point, effecting long-term change requires "multiple organizations of individuals as well as institutions, both separately and collaboratively, work[ing] to support and sustain transformation beyond institutional practices" (Clemens 1993, 792).

In this last use of transformative leadership, one can also see the first sense of the term: the idea of the leader engaged in a collaborative endeavor, one who is embedded in the community which is often diverse, with multiple, even conflicting interests, and one who adds values and vision to the role of leading (Marcus 1993; Boyte and Kari 1996).[23] As the original WELI proposal states:

> Transformative leadership is founded on the concept of collaborative leadership, i.e., leadership that is shared among members of an organization. . .It emphasizes the importance of trust, open communication, shared vision, and shared power. . .[Its] core is a commitment to enhancing the human capital of another for the advancement of a common cause. It is a model of stewardship that can effect change at all levels, from eliminating the micro-climate inequities found in departments to reshaping the broader aspects of institutional culture. [T]he development of a new cadre of inclusive leaders is critical for embedding an institution with values and ethics grounded in cooperation, community, and relationship. . . .Organizations are more likely to thrive within today's complex environments when leadership comes from many places and when it is not confused with formal authority (WEFN 3).

The original "Women Engineering Faculty Leadership Network" saw as its central mission promoting this view of transformative leadership and in developing the skills and "critical connections" between various organizations to support and promote women engineers "both in education and industry" to make it an institutional force and reality (WEFN 3).

The second term, "instigators," comes from Astin and Leland's study (1991) of the very small number of academic women in the highest positions of leadership. By interviewing 35 women leaders, including such women as Donna Shalala, Astin and Leland developed the concept of the "instigator" to describe a cohort group that had a significant impact on higher education. Part of what their study discovered was that these women had built their own success, in part, by developing

---

[22] M. Olson, *The Logic of Collective Action: Public Goods and the Theory of Groups* (Cambridge, MA: Harvard University Press, 1965); E.S. Clemens, "Organizational Repertoires and Institutional Change: Women's Groups and the Transformation of U.S. Politics, 1890–1920," *American Journal of Sociology* 98 (1993): 755–798.

[23] H.D. Boyte and N.N. Kari, *Building America: The Democratic Promise of Public Work* (Philadelphia: Temple University Press, 1996).

formal and informal networks in and outside of their home institutions to deal with their very small numbers. These networks helped them learn from each other (including each other's mistakes) and create influence through this network as a "new broad-based organizational entity" which "served to increase the visibility and viability of individual institutional transformation" (WEFN 3). WELI has, of course, built and attempted to promote "a similar network of instigators among women engineering faculty across the country" and "to bring together people who see the importance of change, who have ideas and who understand the processes and challenges" (WEFN 3).

In many respects, the strength of WELI, carried through its conferences and network, is in developing leaders and, more particularly, in developing transformative leaders—although these results are very difficult to assess. On the one hand, one can see an increased number of women engineering academic leaders, including highly visible leaders, while, on the other, attempts to correlate this increased number with participation in the WELI Conferences would be to overstretch available data. Thus, while no cause–effect comparisons can be made, Table 4.7 indicates the increase in the percentage of women engineers who have achieved a level of leadership (chair and above) over the past five years. Likewise, the change in the number of women engineering faculty within the top

TABLE 4.7: Number of women engineers in academic leadership positions, AAU universities 2000	
• 2.7% (eight women chairs in 298 engineering departments—2000 survey); Niemeier and Gonzalez 2004)	
• 11 women deans (Thompson 2000)	
Number of women engineering professors in top 50 universities, (ASEE, 2007)	
2001	**2006**
• 5.2% Full Professors (*N*=3250)	• 7.9% Full Professors (*N*=3572)
• 15.2% Associate Professors (*N*=1082)	• 16.7% Associate Professors (*N*=1108)
• 23.5% Assistant Professors (*N*=817)	• 28.6% Assistant Professors (*N*=985)
• Overall mean = 10.2%	• Overall mean = 13.2%
Number of women engineers in academic leadership positions, 2006	
• 26 women deans (source: Engineering Deans Council)	
• 20 department chairs (WEPAN 2006)	
• 9 provosts/vice-provosts (WEPAN 2006)	
• 8 presidents/chancellors (WEPAN 2006)	
• 13.8% of doctorate-granting institutions have women presidents (all disciplines) (ACE 2007)	

50 institutions (Table 4.7) in the United States follows the same pattern as increases in administrators. While this is a small difference, there is an increase in women engineering faculty at all ranks: from 5.2% in 2001 to 7.9% in 2006, thus creating a larger pool for leadership.

As mentioned, while these numbers are inspiring, they do not address women science and engineering leaders' reflections on academic organizational culture—specific department or university settings, the general patterns that seem to persist across institutions or institutional types (private, public, research universities, etc.), the challenges of the job, etc. It is not only that, in the case of WELI, numbers are not enough to truly capture and understand women scientists' and engineers' leadership experiences, but that without really understanding why these numbers have improved, they are difficult to sustain, especially when, as we are often told by women colleagues who have taken on leadership roles, the experience of leading can itself prove to be disheartening. Numbers may improve temporarily, for instance, but decline over the long-term; or women may be interested only in brief forays into leadership positions. A related challenge is that women engineers and scientists may, through leadership development programs, be encouraged to take on leadership roles at the expense of their primary research and teaching functions, so that the numbers of leaders may improve while the quality of research among senior women faculty wanes, along with their status as educational role model in the classroom. These are the complications involved which a focus exclusively on quantitative measures cannot address. Likewise, these numbers cannot address the multiple institutional challenges which women leaders face, or their insights into the subtle rules of leadership success. We turn to these next.

CHAPTER 5

From Success Stories to Success Strategies: Leadership for Promoting Diversity in Academic Science and Engineering

5.1   INTRODUCTION: SUCCESS STRATEGIES AND "INSIDER KNOWLEDGE"

The National Research Council's *To Recruit and Advance: Women Students and Faculty in Science and Engineering* (2006, 2) distinguishes between "best practices" and "successful strategies": the latter term "best" presumes a set of fixed or interchangeable practices that will work anywhere regardless of context. We too want to begin this chapter by recognizing that success, when it comes to climate change, is context-specific and variable, that strategies are appropriate and meaningful only in light of a given institution.[1] We also prefer "successful strategies" to "best practices" because this term allows for a dialog across differences and, hence, a more inclusive approach to strategies developed among various programs that may employ different means to achieve shared ends, in this case, fostering the participation and success of advanced women and women leaders in S&E.

In this chapter, we discuss our experiences in developing what amounts to a transformative leadership curriculum for women engineers and scientists at the fourth WELI Conference at Syracuse University in 2004–2005. As mentioned, our understanding of the nature of academic leadership, including the challenges and critical issues for women in S&E, is filtered through the lens of this experience. But we have also synthesized theory with practice, integrated a burgeoning interdisciplinary literature on gender, leadership, organizations, and change with our self-reflexive insights into developing a leadership program, including thinking critically about results. In the first section of this chapter, we detail the process of building a leadership curriculum specific to the needs and challenges

---

[1] National Research Council, *To Recruit and Advance: Women Students and Faculty in Science and Engineering* (Washington, D.C., The National Academies Press, 2006): 2.

of women academic engineers and scientists, a group brought together by three very different research institutions in the Northeast, Canada, and the South. We drill down, not only on the motivations, activities, and outcomes of the *Advanced Leadership & Institutional Transformation Workshop and Conference*, but our special focus on transformative leadership—the interaction between developing leadership capacity and institutional transformation. In doing this we speak of the "successful strategies" that both became a dynamic part of our *Conference* and filtered into the WELI network in general.

The core premise of our particular *Conference*, evident in the curriculum and programming, and which resonates across many women in S&E programs, is that successful strategies for advancing women leaders cannot be garnered from academic research alone, but must be determined by drawing on "insider knowledge"—the vast, divergent experiences of women academic scientists and engineers, including their experiences with leadership (as leaders, as followers) and with some aspect of diversity. The experiences of the diverse group of *Conference* attendees, both speakers and participants, offered invaluable insight into the many aspect of leadership that we have already mentioned: academic organizational culture and the range of department cultures; formal and informal networks; gender and difference as they inflect leadership style; the specific demands expected of women as non-traditional leaders in male-dominated contexts; gender roles and socialization as they impact career advancement; gender and communication, etc.

5.2   WELI ADVANCED LEADERSHIP & INSTITUTIONAL TRANSFORMATION CONFERENCE

Conference attendees comprised over 60 junior, mid-career, and senior level leaders largely from academic institutions, though affiliates from professional organizations, government, and industry were also present. Despite different backgrounds, disciplines, and institutional locations, attendees defined their role as stakeholders in the ADVANCE and WELI mission of supporting women and minority leaders in academia to improve the overall climate in S&E. Likewise, the *Conference* utilized the interactive workshop format to bridge the many gaps: between academic disciplines; between cultures of academia, industry, and professional organizations; and between S&E members and additional stakeholders in implementing diversity programs such as faculty in women's studies, the social sciences, policy initiatives, law, etc. *Conference* organizers carefully selected speakers and participants from underrepresented minority groups in engineering and included past NSF/WELI leaders as speakers and panelists.

5.2.1   Twin Goals: Individual Advancement and Institutional Change

By design, each *Conference* session began with a speaker's forthright discussion of her leadership trajectory—career path, struggles, challenges, experiences, and mistakes. A range of issues were, thus, placed on the table for discussion right away: nonlinear career paths, the arbitrariness of one's discipline choice in graduate school, unsuccessful mentor relationships, feeling underprepared to lead a department, the difficulties of raising children and conducting research, underestimating the importance of communication skills, etc. Taken together, these "introductions" amounted to a rich tapestry of reflections at the interface of individual success and larger patterns of institutional barriers in academic S&E. They also functioned to give audience members permission to speak forthrightly and self-reflexively about their own experiences and to begin to construct a narrative roadmap about their aspirations for career advancement and leadership.

To place the value of this practice in context, some background on the *Conference* is necessary. Our special focus was transformative leadership—the interaction between leaders and institutional transformation—which was itself identified as one critical need area at the early October 2000 WELI Conference. When we examined the results from the three prior WELI Conferences in Alabama (2003), Utah (2003), and Connecticut (2004), all of which emphasized the importance of creating a collaborative vision between speakers and participants, we prioritized twin goals: advancing women's individual and institutional success as academic leaders in S&E. Thus, we attempted to "humanize" institutional transformation so as to keep our sights firmly focused on the women themselves—their needs, experiences, concerns, and visions. Only by doing this, we believed, could institutional transformation become a practical reality and not a burden to individual women. In fact, a common complaint by women academics, both in and beyond S&E, is the demands of service, the fact that women are often disproportionately saddled with doing the administrative work of departments and colleges, advising and mentoring students, running extracurricular programming, etc. The hidden cost of these otherwise very worthwhile investments is that women may find themselves behind in their research performance.

Beyond individual women's success, we also focused on resourcing women leaders for success in the processes of institutional transformation. This included discussions of successful strategies as conveyed by practicing leaders, but also familiarizing participants with the body of research on gender and leadership. In many respects, this research literature provided participants with a baseline reference for evaluating their own career prospects and anticipating challenges, a conceptual language to identity certain institutional issues and potential problem areas, and a space to engage in creating a personal vision of leadership to negotiate these potential pitfalls. A memorable moment, for instance, was the impact among audience members of Tracey Manning's discussion of self-efficacy—different from a general sense of confidence or self-esteem—as the specific belief in one's own ability to succeed and to persevere even under pressures or challenges. Essentially, Manning described how women in non-traditional leadership roles who possessed high leadership self-efficacy were able to deal better with various predictable challenges, including their own self-doubt borne of stereotype threat, others' biased evaluation of their leadership performance, and institutional barriers and issues.[2] These critical concepts of self-efficacy and stereotype threat (i.e., the fear that one's behavior will confirm stereotypes about the group with whom one identifies and the resulting impairment of performance) armed participants with critical tools to analyze their own leadership practices in context. Likewise, centrally important to both individual and institutional goals, as in all the WELI Conferences, was the role of supportive connections via a national network.

In order to create a distinctive *Conference* experience and to integrate our twin objectives in our leadership training and development curriculum, we decided that the *Conference* program would not only have to be collaborative, but it must involve direct and personal input from as many speakers and participants as we could contact. We talked to every speaker and many of our participants, which in many cases involved traveling to cities around the United States for in-depth discussions about leadership experiences, perspectives, future hopes, initiatives, and goals.

5.2.2   Action-Oriented Roadmap

In addition, the *Conference* distinguished itself from prior WELI Conferences by a three-part, action-oriented curriculum on leadership self-assessment, current leadership theory, and developing a career planning roadmap—all in the service of promoting *both* individual success and successful strategies for institutional transformation. The leadership development and training curriculum, for instance, included interactive self-assessment components, joint leadership scholarship and leadership coaching modules, mock-scenarios and simulations of leadership and institutional challenges, and many experience-oriented panels and discussions. Its sources came from two areas: (a) current leadership scholarship, including new literatures on women leaders and academic leadership; and (b) the needs, experiences, and perspectives of individual women S&E leaders in academia.

5.2.3   Mapping Out Individual Goals in Context

From carefully interlinking these two bodies of knowledge and information, we identified four priorities, each embedded in an action-oriented practice:

1. *Engagement in leadership visioning and theorizing:* by emphasizing critical thinking and familiarity with leadership scholarship and studies, women faculty can define and set their own terms for leadership, including leadership style, in often male-dominated climates that they know best

---

[2] Crystal L. Hoyt, Jim Blascovich, "Leadership Efficacy and Women Leaders' Responses to Stereotype Activation," *Group Processes and Intergroup Relations* 10(4)(2007): 598–616; D.M. Bergeron, C.J. Block and B.A. Echtenkamp, Disabling the Able: Stereotype Threat and Women's Work Performance," *Human Performance* 19 (2006): 133–158.

2. *Leadership self-assessment:* by assessing their own individual style through a leadership assessment questionnaire tailored for women scientists and engineers, women faculty can identify and expand on their strengths, develop strategies to anticipate and contend with their weaknesses, and further develop their own situational style of leading specific to their situation—department/university, level of institutional transformation support, etc.

3. *Leadership skill development specific to the challenges and opportunities of a given academic setting:* by examining the institutional context of leadership—including matching the leader to leadership context—women faculty can develop leadership skill sets tailored to the specificities of their environment and its patterns of influence, including addressing such topics as persuasion, communication, budgeting, etc.

4. *Career planning roadmap:* women faculty were able to take time out to plan a career trajectory within the full realization of individual leadership style, studies on leadership, the specific challenges of a given institutional context, and the resources of a supportive network.

Essentially, our main aim was to personally and conceptually arm women faculty with cutting-edge leadership theories and skill sets, including a sense of their own leadership style, to face a range of different climates, including difficult, inhospitable, or impenetrable settings, in their respective contexts.

5.2.4   Strategies: Individual to Institutional Success

If one central premise of the *Conference* was that individual success in leadership, in turn, provides a significant conduit for greater institutional transformation, several strategies were identified to make this possible. They included:

- Helping women leverage their own and others' leadership experiences into a knowledge base helpful for their individual success as leaders
- Using leadership self-assessment and self-reflection to demystify the talents of good leadership and leadership skill sets and success strategies
- Promoting career planning through individual awareness of the strengths and limitations of personal leadership style in relation to institutional context
- Recognizing and articulating the links between an individual leader, a community of women leaders in S&E, and transforming academic engineering and science

5.3   ASSESSMENT RESULTS: WELI ADVANCED LEADERSHIP & INSTITUTIONAL TRANSFORMATION CONFERENCE

Tables 5.1 and 5.2 detail the make-up of *Conference* participants and the topics, themes, and approaches developed at the *Conference*. Not only did participants represent a range of fields in engineering, as Table 5.1 shows, they came from a range of ranks.

TABLE 5.1: Participants by field and rank		
FIELD	**FREQUENCY**	**PERCENT**
Biomedical Engineering	2	6.7
Chemical Engineering	7	23.3
Civil and Environmental Engineering	6	16.7
Computer Engineering/Science	3	10.0
Electrical Engineering	6	20.0
Industrial Engineering	2	6.7
Materials Science and Engineering	1	3.3
Mechanical Engineering	3	10.0

RANK	**FREQUENCY**	**PERCENT**
Dean	3	10.0
Associate Dean	2	6.7
Department Chair	7	23.3
Director	3	10.0
Professor	8	26.7
Associate Professor	6	20.0

Some general "lessons learned" from pre-conference surveys include:

- Few (31%) participants reported that they have mentors to assist them in career advancement, while the majority indicated inadequate resources for career planning and developing leadership opportunities.
- Most participants held positional leadership roles (87%) and most considered themselves leaders (75%).
- Most participants expressed a need for learning more about the types of leadership roles and positions in academia and the range of skills that they would need to be successful in these positions.

TABLE 5.2: Conference content and approaches
Contrasting perspectives on good leadership and effective leadership style
Leadership from the vantage point of institutional position (including academic chairs, research directors, deans, and non-titled positions) and context (liberal arts school, research university, geographical/regional differences)
Challenges for the advanced leader as compared to beginning an administrative career
Leadership skill sets
Leadership scholarship on gender, academia, and in the sciences and engineering
Women academic leaders in engineering as "change agents"
Diversity and leadership style
"Matching" leadership style with institutional situation
Simulating mock scenarios to build effective leadership strategies
Case Studies approach
Meeting institutional challenges
Crafting a personal leadership vision
Drafting a career action plan
Institutional responsibilities such as fundraising, diversity, and representing minority groups
Theorizing the next steps for women in academic leadership in engineering

- Most participants ranked "good communication" as the most important criteria for "good leadership."
- Few (19%) participants felt that they had "adequate resources" and a "plan for meeting leadership objectives," while the majority felt that they had neither resources nor a career plan for developing leadership opportunities.
- While the majority of attendees held positional leadership roles (87%), such as dean or department chair, 13% reported that they were influential on their campus in a non-positional role (e.g., leadership outside the normal administrative roles, including being recognized as an outstanding teacher or researcher).

- Many (69%) attendees had participated in previous leadership training, but still did not feel adequately resourced; participants described that this prior training was aimed at improving communication skills (22%), developing negotiation skills (22%), and/or developing networks (56%).

- When asked about their personal leadership goals and aspirations in preconference surveys, a little less than half (6 of 15 who answered this question) indicated that they were unsure.

- Pre-conference evaluations also queried participants on what they deemed the three most important characteristics of good leadership, which are summarized in Table 4.3 in order of the most responses received.

- In post-conference evaluations of the panels and events, most participants (22 of 29) indicated that the leadership trainings were most helpful or helpful.

5.4   SUCCESS-ORIENTED PRINCIPLES FOR DIVERSITY LEADERSHIP

There are many contributions made by WELI: the introduction of key themes across the WELI Conferences; suggested successful strategies; training modules; and leadership development pedagogies for women scientists and engineers, etc. One can also appeal to the writings and results from each of the WELI Conferences, i.e., program evaluations, discussions with participants, or discoveries implemented in participant's home institutions (Anderson-Rowland et al. 2006; Cooper 2004; Davidson, Vance, and Niemeier 2002; Layne 2004; Lighty, Vance and Niemeier 2005; Perusek 2004). We have already described many of these insights in the preceding chapter. An additional important contribution was in our attempt to make attendees' time at the *Conference* indisputably worthwhile by offering many opportunities for networking, peer-engagement, and personally connecting with others. This enabled participants to be part of a team in working collaboratively with others at the *Conference* and in ongoing ways in the future, including engaging in a synthetic curriculum; and prioritizing usable tools to be taken back to participants' respective campuses.

However, in the process of defining our *Conference* objectives, we identified several propositions that informed, not only the WELI leadership mission across the many Conferences, but a set of underlying premises and principles relevant for strengthening the university as a mission-driven institution in general. We have found that these propositions, listed below, function as a kind of analytical vision with which to develop and sustain leadership programs for underrepresented groups in academia in other contexts in the future. From this analytical vision, one can then define and describe future areas of research.

5.4.1   Principles for Diversity Leadership in the Academic S&E Context

These propositions and their underlying principles include:

1. *A diverse constituency is essential to the future of the S&E professions, not only for meeting new demographic realities, or for maintaining U.S. leadership in the technical fields in the global context, but for retaining a sense of the legitimacy of the S&E fields.*

Projections of U.S. workforce needs over the next two decades indicate a significant increase in the demographic diversity of the workforce, a gap in available scientists and engineers in meeting those needs, which may imperil a U.S. global competitive advantage, and, thus, the need to leverage underrepresented groups for the S&E professions. But in addition to these pragmatic issues, these professions are strengthened, not only when their composition represents the diversity of U.S. society, but when they reflect the diverse range of experiences, perspectives, and ideas that characterize this society.

2. *A network of diverse leaders strengthens S&E and academic institutions more generally.*

Diversity is not only a functional goal, but a core value at the heart of the academic and intellectual mission of the college/university and a shared value of the S&E fields. Leaders play a crucial role—as advanced mentors, role models, change agents, institutional figureheads, and power brokers—in fostering diversity and culture change *en route* to bringing S&E into accordance with the full diversity of contemporary society. But because minority leaders remain relatively isolated in these often homogenous fields, a well-established network is needed to encourage, recruit, and train diverse leaders and support their efforts toward institutional change.

3. *Academic leadership plays a critical role in motivating change, in improving and defining the unique organizational culture of the college/university, and in fostering diversity in S&E beyond the academic context in lasting ways.*

The college/university setting offers unique challenges for leaders in general, with the traditionally male-dominated S&E fields furthering those challenges. This setting is characterized both by extremely low numbers of women and diverse leaders in S&E, even while it functions as a key mechanism to influence the constituency of the S&E professions more generally. Intervening in this setting by advancing women and diverse leaders, thus, impacts both the college/university setting and the S&E professions at large.

4. *Women leaders attract more women and promote diversity at all levels of S&E.*

Since women and minorities often do not see S&E as an appropriate or inviting profession, or have little first-hand knowledge of these fields, women and diverse leaders are particularly important to change the "face" of the S&E fields and professions, to attract young women and underrepresented men to S&E, and to reform the environment in such a way that expands the definition and practice of S&E. This reform includes changing cultural representations of the scientist and engineer.

5. *Advanced academic women leaders need targeted strategies, discussions, coordinated team efforts, research initiatives, and institutional resources to achieve greater overall lasting benefits in these challenging environments.*

While there is much research on leadership for success in business management and political governance models, and while women leaders have become an important subject of study in private and public contexts, women academic leaders—especially at advanced levels and in the S&E fields—are understudied in both academic research and in practical leadership development circles. To achieve success in these challenging environments, coordinated efforts and institutional commitments are necessary, as are further inquiries into the nature of leadership for this particular group and its institutional and social impact.

6. *Leadership for transforming institutional climates must include diverse voices, based on a broad definition of diversity: i.e., including gender, ethnic, cultural, and regional backgrounds; academic field specialties; institutional and leadership positions and styles, etc.*

A range of leadership experiences and perspectives, including reflections on failures and mistakes, contrasting viewpoints, and various leadership positions, are necessary to foster a supportive and genuine rapport in leadership inquiry and practice.

7. *Since exposure over time has been shown to increase the positive perception of women leaders' performance and effectiveness, women leaders in academic S&E must develop personal and institutional strategies that enhance self-efficacy, persistence, and endurance.*

Recent compelling research indicates that exposure over time is a significant factor in increasing the positive perception of the performance of women leaders, especially in traditional or male-dominated contexts. In this respect, women academic leaders in S&E need to plan and deliberately develop personal and institutional resources to help them persist and endure in leadership positions.

8. *Women academic leaders in S&E invariably fit into and expand the larger picture of research on women leaders.*

Validating continued inquiry into how women leaders in S&E represent an important area of experience for broader studies of academic and "diversity leadership" is critically important. Women leaders in S&E represent part of a burgeoning field of non-profit and academic leadership and its inquiry into new organizational cultures, the structure of leader–follower relations, and different rules and styles of leadership. Moreover, this group can play an important role in challenging implicit leadership theories—culturally shared perceptions and stereotypes about who is ostensibly "naturally" fit to lead.

9. *The benefits of transformative leadership and institutional transformation outweigh the difficulties of achieving benchmarks in these areas.*

While research has shown that it is much harder to lead diverse groups, it is also the case that diverse groups offer more creative approaches to given issues. Moreover, transformational leaders help produce better quality results and more inclusive environments.

10. *Successful change-oriented leadership is attentive to the continuities between local (department) culture, university mission, and broader social and cultural change.*

Women academic leaders in S&E have a practical and cultural-representational role to play at the intersection of departments, universities, and the culture at large.

5.4.2   Future Research

Future research directions for women academic leaders in the S&E must emphasize processes for assessing successful innovations delivered by change agents/transformative leaders; new theories of non-traditional leadership style in relation to traditional organizational culture; strategies for combating cultural stasis in organizational culture; balancing the multiple institutional challenges that women academic leaders are faced with such as achieving parity, inclusivity, and progress while navigating the inertia and socialization of certain climates.

CHAPTER 6

Conclusion

Over the last several decades, women have made an impressive incursion into S&E, especially in education. In many respects, this feat—begun with broad-based social and cultural changes during the Civil Rights Movement of the 1960s and 1970s—has succeeded in raising new issues in S&E to a national and international level of visibility, including new roles for women.

One increasingly important area in this context in which women are currently making change is in academic leadership. In part, it is due to necessity. The social landscape in which women are intervening is one of severe underrepresentation in both the S&E professional work-force and in advanced academic positions. Women, for instance, comprise nearly half of those with college degrees in the workforce, but make less than 25% of the S&E component of that workforce. African Americans, Latinos, and Native Americans, too, participate in the S&E work-force at rates far below their role in the overall workforce (BEST *Report* 2004b). In the academic context, women's representation begins with promising numbers in 2005 in S&E, earning half of all bachelor's degrees, 44% of all master's degrees, and 45% of all doctorates, but women ulti-mately represented less than half of the male doctoral faculty (60,000 men compared with 25, 000 women) employed in Research I institutions (for 2003) (NSF 2007). The consequences of this discrepancy are significant: isolation of women, lack of role models and mentors for the next gen-eration of women scientists and engineers, gendered segregation of the workforce, narrow cultural representations of the scientist and engineer and the nature of their work, continued social and institutional bias and barriers, unreformed academic governance and reward policies, a national shortage of scientists and engineers, skewed composition of the S&E workforce as compared with the future U.S. population, problems in a U.S. global competitive edge and resource and develop-ment allocation, etc.

To redress this situation, women academics in S&E have begun to prioritize institutional transformation through the mechanism of academic leadership. This endeavor has meant that women faculty in the S&E fields are taking an activist role in confronting the persistent problems and obstacles that remain as barriers to women's full participation in academic S&E and creating a range of solutions to help diverse constituencies advance and succeed in the technical fields in and beyond the academic context. Most critically, in this role of making change, women academic scientists and engineers have prioritized individual and institutional goals, infused new energy into leadership research, practice, and program development, and given shape to some of the most pertinent challenges of diversifying S&E education faced by colleges/universities now and in the near future.

·   ·   ·   ·

References

The 2002 Global 500: The CEOs. 2002. *Fortune.*

Academe Online. 2004. Balancing faculty careers and family work. 90(6).

Acker, J. 1991. Hierarchies, jobs, bodies: A theory of gendered organizations. In S. Farrell, ed. *The social construction of gender*, 162–179. Newbury Park, CA: Sage.

Adler, N.J. 1999. Global leaders: Women of influence. In G.N. Powell, ed. *Handbook of gender and work.* Sage, Thousand Oaks, CA, 239–261.

Akerlof, G. and R. Kranton. 2000. Identity and economics. *The Quarterly Journal of Economics* 115.3: 715–754. doi:10.1162/003355300554881

Aguirre, Adalberto. 2000. Academic storytelling: A critical race theory story of affirmative action. *Sociological Perspectives* 43.2 (Summer) 319–339.

Aguirre, A. and R. Martinez. 2002. Leadership practices and diversity in higher education: Transitional and transformational frameworks. *Journal of Leadership Studies* 8.3: 53.

Aldrich, H. 1989. Networking among women entrepreneurs. In 0. Hagan, C. Rivchun and D. Sexton, eds. *Women-owned businesses:* 103–132. New York: Praeger.

Aldrich, H., P.R. Reese, and P. Dubini. 1989. Women on the verge of a breakthrough: Networking among entrepreneurs in the United States and Italy. *Entrepreneurship and regional development.* 1: 339–356. doi:10.1080/08985628900000029

American Association for the Advancement of Science (AAAS). 2001. *In pursuit of a diverse science, technology, engineering, and mathematics workforce: Recommended research priorities to enhance participation by underrepresented minorities.* Y.S. George, D.S. Neale, V. Van Horne, and S.M. Malcom. Washington, D.C.: AAAS

American Council on Education. 2005. *An agenda for excellence: Creating flexibility in tenure-track faculty careers.* Washington, DC: ACE.

Anderson-Rowland, M., B. Homsher, J. Lighty, J. Raper, and J. Vance. 2006. A characterization of potential women engineering administrators in academia. *WEPAN 2006 National Conference Proceedings.*

Antonio, A. L., M. J. Chang, K. Hakuta, D. A. Kenny, S. Levin, and J. F. Milem. 2004. Effects of racial diversity on complex thinking in college students. *Psychological Science*, 15 (8): 507–510. doi:10.1111/j.0956-7976.2004.00710.x

Arenson, Karen W. 2002. More women taking leadership roles at colleges. *New York Times*. 4 July.

Aronson, J. and C.M. Steele. 2005. Stereotypes and the fragility of human competence, motivation, and self-concept. In C. Dweck and E. Elliot, eds., *Handbook of competence and motivation*. New York: Guilford.

Askling, Berit and Bjørn Stensaker. 2002. Academic leadership: Prescriptions, practices and para-doxes. *Tertiary Education and Management* 8: 113–125.

Astin, A.W. 1993. Diversity and multiculturalism on the campus: How are students affected? *Change* (March/April)

Astin, H.S. 2004. Women as transformational leaders. Paper presented at the American Psycho-logical Association, Honolulu, HI (July).

Astin, A.W. and H. Astin. 2000. *Leadership reconsidered: Engaging higher education in social change*. Battle Creek, MI: W.K. Kellogg Foundation.

Astin, Helen S. and Carole Leland, eds. 1991. *Women of influence, women of vision: A cross-generational study of leaders and social change*. San Francisco: Jossey Bass.

Arrow, K. 1973. The theory of discrimination. In O. Ashenfelter and A. Rees, eds., *Discrimination in labor markets*. Princeton, NJ: Princeton University Press: 3–33.

Avolio, B. J. Sosik, D.I. Jung and Y. Berson. 2003. Leadership models, methods, and applications. In D. K. Freedheim, I.B. Weiner, W.F. Velicer, J.A. Schinka, R.M. Lerner, eds. *Handbook of psychology* (V12): 277–307. Hoboken, NJ: Wiley. doi:10.1002/0471264385.wei1212

Bailyn, L. and J.K. Fletcher. 2003. *The equity imperative: Reaching effectiveness through the dual agenda* (Briefing Paper #18). Boston: Center for Gender in Organizations, Simmons School of Management.

Bajdo, Linda M. and Marcus W. Dickson. 2001. Perceptions of organizational culture and wom-en's advancement in organizations: A cross-cultural examination. *Sex Roles* 45.5–6 (Sept) 399–414.

Bakken, L.L. 2005. Who are physician-scientists' role models? Gender makes a difference. *Academic Medicine* 80.5 (May): 502–506. doi:10.1097/00001888-200505000-00020

Balkundi, Prasad and Martin Kilduff. 2006. The ties that lead: A social network approach to leader-ship. *The Leadership Quarterly* 17: 419–439. doi:10.1016/j.leaqua.2006.01.001

Bartol, K.M. and D.C. Martin. 1986. Women and men in task groups. In R.D. Ashmore and E K. Del Boca, eds. *The social psychology of female–male relation: A critical analysis of central con-cepts*. Orlando, FL: Academic Press: 259–310.

Bass, B. and R.M. Stogdill. 1990. *Bass and Stogdill's handbook of leadership: Theory, research, and managerial applications*. New York: Free Press.

Bass, B. 1990. From transactional to transformational leadership: Learning to share the vision. *Organizational Dynamics* (Winter): 19–31. doi:10.1016/0090-2616(90)90061-S

Bass, B. and B. Avolio, eds. 1993. *Improving organizational effectiveness through transformational leadership.* Newbury Park, CA: Sage.

Baumeister, R.F., and M.R. Leary. 1995. The need to belong: Desire for interpersonal attachments as a fundamental human motivation. *Psychological Bulletin,* 117: 497–529. doi:10.1037/0033-2909.117.3.497

Beaman, Lori, Raghabendra Chattopadhyay, Esther Duflo, Rohini Pande and Petia Topalova. 2008. Powerful women: Does exposure reduce prejudice?. Weatherhead Center for International Affairs, Harvard University: WCFIA Working Paper. http://econ-www.mit.edu/files/2406.

Becker, G. 1957. *The economics of discrimination.* Chicago: Chicago University Press.

Bell, Myrtle, Mary McLaughlin, and Jennifer Sequeira. 2002. Discrimination, harassment, and the glass ceiling: Women executives as change agents. *Journal of Business Ethics* 37.1 (April): 65–76.

Bem, S.L. 1981. Gender schema theory: A cognitive account of sex typing. *Psychological Review* 88: 354–364. doi:10.1037//0033-295X.88.4.354

Bennis, W. and B. Nanus. 1985. *Leaders: The strategies for taking charge.* New York: Harper and Row.

Beoku-Betts, J. 2004. African women pursuing graduate studies in the sciences: Racism, gender bias, and Third World marginality. *NWSA Journal* 16.1: 116–135. doi:10.2979/NWS.2004.16.1.116

The BEST Initiative. 2004b. *The talent imperative: Meeting America's challenge in science and engineering.* Building Engineering and Science Talent, San Diego, CA. http://www.bestworkforce.org/PDFdocs/BESTTalentImperativeFINAL.pdf.

The BEST Initiative. 2004a. *A bridge for all: Higher education design principles to broaden participation in science, technology, engineering and mathematics.* Building Engineering and Science Talent, San Diego, CA. http://www.bestworkforce.org/PDFdocs/BEST_BridgeforAll_HighEdFINAL.pdf

Biernat, Monica and Kathleen Fuegen. 2001. Shifting standards and the evaluation of competence: Complexity in gender-based judgment and decision making. *Journal of Social Issues* 57(4): 707–724. doi:10.1111/0022-4537.00237

Biernat, M. and D. Kobrynowicz. 1997. Gender- and race-based standards of competence: Lower minimum standards but higher ability standards for devalued groups. *Journal of Personality and Social Psychology* 72: 544–557 doi:10.1037/0022-3514.72.3.544

Birnbaum, Robert. 1992. *How academic leadership works: Understanding success and failure in the college presidency*. San Francisco, CA: Jossey-Bass.

Bix, A.S. 2004. From 'Engineeress' to 'Girl Engineers' to 'Good Engineers': A history of women's U.S. engineering education. *NWSA Journal 16*.1: 27–49. doi:10.2979/NWS.2004.16.1.27

Blackmore, Paul. 2007. Disciplinary difference in academic leadership and management and its development: a significant factor? *Research in Post-Compulsory Education*, 12.2 (July): 225–239. doi:10.1080/13596740701387502

Bleier, Ruth. 1998. The cultural price of social exclusion: Gender and science. *NWSA Journal* 1 (Autumn 1988): 7–19.

Bombardieri, Marcella. 2005. Summers' remarks on women draw fire. *Boston Globe*, 17 Jan.

Book, E.W. 2000. *Why the best man for the job is a woman*. New York: Harper Collins.

Bosson, J.K., E.L. Haymovitz and E.C. Pinel. 2004. When saying and doing diverge: The effects of stereotype threat on self-reported versus nonverbal anxiety. *Journal of Experimental Social Psychology* 40: 247–255. doi:10.1016/S0022-1031(03)00099-4

Bosson, J.K., J.N. Taylor, and J.L. Prewitt-Freilino. 2006. Gender role violations and identity misclassification: The roles of audience and actor variables. *Sex Roles*. 55(1–2): 13–24. doi:10.1007/s11199-006-9056-5

Boyte, H.D. and N.N. Kari. 1996. *Building America: The democratic promise of public work*. Philadelphia: Temple University Press.

Brasileiro, A.M. ed. 1996. *Women's leadership in a changing world*. Washington, D.C.: UNICEM.

Brass, Daniel J., Joseph Galaskiewicz, Henrich R. Greve, Wenpin Tsai, 2004. Taking stock of networks and organizations: A multilevel perspective. *Academy of Management Journal* 47.6: 795–817.

Brown, F. William and Dan Moshavi. 2002. Herding academic cats: Faculty reaction to transformational and contingent reward leadership by department chairs. *Journal of Leadership & Organizational Studies* 8(3): 79–95. doi:10.1177/107179190200800307

Buckingham, M. 2005. What great managers do. *Harvard Business Review* 83.3, March.

Bureau of Labor Statistics, U.S. Department of Labor, 2002, Household Data: Annual Averages, Table 11: Employed Persons By Detailed Occupation, Sex, Race, And Hispanic Origin, www.bls.gov/cps/cpsaat11.pdf.

Burgess, D. and E. Borgida. 1999. Who women are, who women should be: Descriptive and prescriptive gender stereotyping in sex discrimination. *Psychology, Public Policy, and Law* 5: 665–692.

Burns, James MacGregor. 2004. *Transforming leadership: A new pursuit of happiness*. New York: Grove Press.

Burns, James MacGregor. 1978. *Leadership*. New York: Harper & Row, Publishers.

Bystydzienski, Jill. 2004. (Re)Gendering science fields: Transforming academic science and engineering. *NWSA Journal* 16.1: viii–xii.

Cain, J.M. et al. 2001. Effects of perceptions and mentorship on pursuing a career in academic medicine in obstetrics and gynecology. *Academic Medicine* 76.6 (Jun): 628–634. doi:10.1097/00001888-200106000-00015

Campbell, K. E. 1988. Gender differences in job-related networks. *Work and Occupations* 15: 179–200. doi:10.1177/0730888488015002003

Campbell, K.E., P. Marsden and J. Hurlbert. 1986. Social resources and socioeconomic status. *Social Networks* 8: 97–117. doi:10.1016/S0378-8733(86)80017-X

Campbell, K.E. and R.A. Rosenfeld. 1985. Job search and job mobility: Sex and race differences. *Research in the Sociology of Work* 3: 147–174.

Canary, Daniel J. and Kathryn Dindia, eds. 1998. *Sex differences and similarities in communication.* London: Lawrence Erlbaum Associates

Carli, Linda L. and Alice H. Eagly. 2001. Gender, hierarchy, and leadership: an introduction. *Journal of Social Issues* 57: 629–636. doi:10.1111/0022-4537.00232

Carli, Linda L., Suzanne J. LaFleur, and Christopher C. Loeber. 1995. Nonverbal behavior, gender, and influence. *Journal of Personality and Social Psychology* 68, 6 (June): 1030–1041. doi:10.1037//0022-3514.68.6.1030

Carli, Linda. Gender and social influence. 2001. *Journal of Social Issues* 57(4): 725–741. doi:10.1111/0022-4537.00238

Carli, Linda and Alice H. Eagly. 2001. Gender, hierarchy, and leadership: An introduction. *Journal of Social Issues* 57: 629–636. doi:10.1111/0022-4537.00232

Carnes, Molly. 1996. Just this side of the glass ceiling. *Journal of Women's Health* 5(4): 283–286.

Carr, Phyllis L. et al. 2003. A 'ton of feathers': Gender discrimination in academic medical careers and how to manage it. *Journal of Women's Health* 12.10 (Nov): 1009–1018. doi:10.1089/154099903322643938

Cavallo, Kathleen. 2006. Emotional competence and leadership excellence at Johnson & Johnson: The emotional intelligence and leadership study. *Europe's Journal of Psychology*, 11 Feb.

Ceci, Stephen J. and Wendy M. Williams, eds. 2006. *Why aren't more women in science? Top researchers debate the evidence.* Washington D.C., American Psychological Association.

Center for the American Woman and Politics. 2002. *Women in Elected Office 2008.* Rutgers-The State University of New Jersey, http://www.rci.rutgers.edu/~cawp/Facts.html#elective.

Chait, R.P., W.P. Ryan, and B.E. Taylor. 2005. *Governance as leadership: Reframing the work of nonprofit boards.* New York: John Wiley & Sons.

Chesler, Naomi C. and Mark A. Chesler. 2002. Gender-informed mentoring strategies for women engineering scholars: On establishing a caring community. *Journal of Engineering Education* 91.1 (Jan): 49–55.

Chamberlain, Mariam. 2001. Women and leadership in higher education. In Cynthia. B. Costello and Anne J. Stone, eds. *The American woman 2001–2002: Getting to the top*, 63–82. New York: W.W. Norton. doi:10.1111/j.1468-0262.2004.00539.x

Chattopadhyay, Raghabendra and Esther Duflo. 2004. Women as policy makers: Evidence from a randomized policy experiment in India. *Econometrica* 72.5: 1409–1443.

Chemers, M.M. 1997. *An integrative theory of leadership*. Mahwah, NJ: Lawrence Erlbaum.

Chliwniak, Luba. 1997. *Higher education leadership: Analyzing the gender gap*. Washington, D.C.: ASHE-ERIC Higher Education Report ED 410 847, 25-4.

Choi, C.Q. 2004. Women scientists face problems. *The Scientist* 18.3 (Feb).

Christensen T. and P. Lægreid. 2001a. A transformative perspective on administrative reforms. In T. Christensen and P. Lægreid, eds., *New public management: The transformation of ideas and practice*. Aldershot: Ashgate.

A Classification of Institutions of Higher Education—1994 Edition. 1994. Princeton, NJ: The Carnegie Foundation for the Advancement of Teaching.

Claes, Marie-Therese. 1999. Women, men and management styles. *International Labour Review*, 138.4: 41–46.

Clemens, Elisabeth S. 1993. "Organizational repertoires and institutional change: Women's groups and the transformation of U.S. politics, 1890–1920." *American Journal of Sociology* 98: 755–98.

Clewell B.C. and P.B. Campbell. 2002. Taking stock: Where we've been, where we are, where we're going. *Journal of Women and Minorities in Science and Engineering* 8.3&4: 255–284.

Collins, Patricial Hill. 1999. Moving beyond gender: Intersectionality and scientific knowledge. In Myra Marx Ferree, Judith Lorber, and Beth B. Hess, eds. *Revisioning gender*. Thousand Oaks, CA: Sage: 261–284.

Colwell, R.R. 2001. Barriers to and opportunities for women in science. Speech to Washington College. Arlington, Virginia: National Science Foundation. www.nsf.gov/news/speeches/colwell/rc011017washcollege.htm.

Congressional Commission on the Advancement of Women and Minorities in Science, Engineering, and Technology Development (CAWMSET). 2004. *Land of plenty: Diversity as America's competitive edge in science, engineering, and technology, http://www.nsf.gov/pubs/2000/cawmset0409/cawmset_0409.pdf*.

Cooper, Nan R. 2004. UConn Engineering hosts Women in Engineering Leadership Summit. *Frontiers, University of Connecticut* (Summer) 12.

Coughlin, Linda, Ellen Wingard, and Keith Hollihan. 2005. *Enlightened power: How women are transforming the practice of leadership*. San Francisco, CA: Jossey-Bass.

Cyert, R.M. Defining leadership and explicating the process. 1990. *Nonprofit management and leadership* 1.1: 29–38.

Davidson, M. and C.L. Cooper. 1992. *Shattering the glass ceiling: The woman manager*. London: Paul Chapman.

Daly, Brenda. 2004. Introduction: *Special Issue:* (Re)Gendering Science Fields. *NWSA Journal* 16.1 (Spring) vi–vii.

Davidson, Valerie. 2001. Women in Engineering Leadership Institute (WELI). Presentation at Second MacKay-Lassonde Memorial Forum, School of Engineering, University of Guelph, Guelph, ON N1G 2W1.

Davidson, V.J., J. Vance and D. Niemeier. 2002. Women in Engineering Leadership Institute: Academic leadership development plans. *ICWES12: 12th International Conference of Women Engineers and Scientists Proceedings*.

Davenport, D.S., and J.M. Yurich. 1991. Multicultural gender issues. *Journal of Counseling and Development* 70.1: 64–71.

Davies, Paul G., Steven J. Spencer, and Claude M. Steele. 2005. Clearing the air: Identity safety moderates the effects of stereotype threat on women's leadership aspirations. *Journal of Personality and Social Psychology* 88(2): 276–287. doi:10.1037/0022-3514.88.2.276

Dean, Cornelia. 2006. Women in science: The battle moves to the trenches. *New York Times*, 19 Dec.

Dean, Cornelia. 2006. Bias is hurting women in science, panel reports. *New York Times*, 19 Sept.

Deaux, K. and L.L. Lewis. 1983. Components of gender stereotypes. *Psychological Documents*, 13.25: Ms. No. 2583.

Diekman, Amanda B. and Wind Goodfriend. 2006. Rolling with the changes: A role congruity perspective on gender norms. *Psychology of Women Quarterly* 30 (4): 369–383.

Ding, Cody, Kin Sing, and Lloyd Richardson. 2007. Do mathematical gender differences continue? A longitudinal study of gender difference and excellence in mathematics performance in the U.S. *Educational Studies: Journal of the American Educational Studies Association*, 40.3: 279–295. doi:10.1111/j.1471-6402.2006.00312.x

DiTomaso, N., D.E. Thompson and D.H. Blake. 1988. Corporate perspectives on the advancement of minority managers. In D.E. Thompson and N. DiTomaso, eds. *Ensuring minority success in corporate management*. New York: Plenum Press: 119–136.

Doré, Butler and F.L. Geis. 1990. Nonverbal affect responses to male and female leaders: Implications for leadership evaluations. *Journal of Personality and Social Psychology* 58: 48–59.

Downey, James. 2001. Guest Editor's Introduction: Academic leadership and organizational change. *Innovative Higher Education* 25(4) Summer.

Dreher, George F. 2003. Breaking the glass ceiling: The effects of sex ratios and work-life programs on female leadership at the top. *Human Relations* 56(5): 541–562.

Duke, D. L., 1998. The normative context of organizational leadership. *Educational Administration Quarterly* 34.2: 165–195. doi:10.1177/0018726703056005002

Duehr, E.E. and J.E. Bono. 2006. Men, women, and managers: Are stereotypes finally changing? *Personnel Psychology* 59.4: 815–846.

Eagly, A.H. and B.T. Johnson. 1990. Gender and leadership style: A meta-analysis. *Psychological Bulletin*, 108(2), 233–256. doi:10.1111/j.1744-6570.2006.00055.x

Eagly, A.H., and S.J. Karau. 2002. Role congruity theory of prejudice toward female leaders. *Psychological Review* 109(3): 573–598. doi:10.1037/0033-2909.108.2.233

Eagly, A.H., and L.L. Carli. 2003. The female leadership advantage: An evaluation of the evidence. *The Leadership Quarterly* 14: 807–834.

Eagly, A.H., and L.L. Carli. 2007. Through the labyrinth: The truth about how women become leaders. Boston: *Harvard Business School Press*. doi:10.1016/j.leaqua.2003.09.004

Eagly, A. H., and A.B. Diekman. 2005. What is the problem? Prejudice as an attitude-in-context. In J.F. Dovidio, P. Glick, and L. A. Rudman, eds. *On the nature of prejudice*: *Fifty years after Allport*, 19–35. Malden, MA: Blackwell.

Eagly, Alice H., Mary C. Johannesen-Schmidt, and Marloes L. van Engen. 2003. Transformational, transactional, and laissez-faire leadership styles: A meta-analysis comparing women and men. *Psychological Bulletin* 129.4: 569–591. doi:10.1002/9780470773963.ch2

Eagly, A.H. and M.C. Johannesen-Schmidt. 2001. The leadership styles of women and men. *Journal of Social Issues* 57.4: 781–797. doi:10.1037/0033-2909.129.4.569

Eckel, Catherine and Philip Grossman. 1998. Are women less selfish than men?: Evidence from dictator experiments.' *Economic Journal* 108(448), 726–735. doi:10.1111/0022-4537.00241

The Edge. 2005. The science of gender and science, the Pinker vs. Spelke Debate. 16 May, www.edge.org/3rd_culture/debate05/debate05_index.html.

Eggins, H. 2000. *Women as leaders and managers in higher education*. Chicago, IL: Open Press.

Eisler, R. 1987. *The chalice and the blade*. Harper San Francisco.

Eisler, Raine. 1997. The hidden subtext for sustainable change. In Willis Harman and Maya Porter, eds. *The new business of business: Sharing responsibility for a positive global future*. San Francisco, CA, Berrett-Koeler/World Business Academy.

Ely, Robin J. 1995. The power in demography: Women's social constructions of gender identity at work. *Academy of Management Journal* 38(3):589–634.

Ely, R. J., and D.D. Meyerson. 2000. Advancing gender equity in organizations: The challenge and importance of maintaining a gender narrative. *Organization* 7(4): 589–608. doi:10.2307/256740

Etzkowitz, Henry, Carol Kemelgor and Brian Uzzi. 2000. *Athena unbound: The advancement of women in science and technology.* Cambridge, MA: Cambridge University Press. doi:10.117 7/135050840074005

Etzkowitz, H., C. Kemelgor, M. Neuschatz, B. Uzzi, B. and J. Alonzo. 1994. The paradox of critical mass for women in science. *Science* 266(5182): 51–54.

European Commission. 2006. *Women in science and technology: The business perspective.* Directorate General for Research. Science and Society; Women and Science. B-1049 Brussels. http://ec.europa.eu/research/science-society/pdf/wist_report_final_en.pdf. doi:10.1126/science .7939644

Fausto-Sterling, Ann. 1992. Building two-way streets: The case of feminism and science. *NWSA Journal* 4.3: 336–349.

Fels, Anna. 2004. Do women lack ambition? *Harvard Business Review* 82(4): 50–60.

Felder, R.M., S.S. Sheppard, and K.A. Smith. 2005. A new journal for a field in transition. *Journal of Engineering Education* 94.1 (Jan): 7–9.

Ferber, Marianne A. and Jane W. Loeb, eds. 1997. *Academic couples: Problems and promises.* Urbana, IL: University of Chicago Press.

Fisher, J. L., and J.V. Koch. 1996. *Presidential leadership: Making a difference.* Phoenix: Oryx Press.

Fiedler, F.E., M. Chemers and L. Mahar. 1978. *Improving leadership effectiveness: The leader match concept.* New York: Wiley.

Fine, Marlene. 2007. Women, collaboration, and social change: An ethics-based model of leadership. In Jean Lau Chin, Bernice Lott, Joy Rice, Janice Sanches-Hucles, eds. *Women and leadership: Transforming visions and diverse voices.* London: Blackwell Publishing: 177–191

Fletcher, J. K. 2001. *Disappearing acts: Gender, power, and relational practice at work.* Boston, MA: MIT Press. doi:10.1002/9780470692332.ch8

Fletcher, J.K. 1998. *Looking below the surface: The gendered nature of organizations* (Briefing Paper, No 2). Boston: Center for gender in Organizations, Simmons Graduate School of Management.

Fletcher, J.K. 1999. *Disappearing acts: Gender, power, and relational practice at work.* Cambridge, Mass: The MIT Press.

Fletcher, J.K. 2003. *The paradox of post heroic leadership: Gender matters* (Working Paper, No. 17). Boston: Center for Gender in Organizations, Simmons Graduate School of Management.

Flynn, Francis and Daniel Ames. 2006. What's good for the goose may not be as good for the gander: The benefits of self-monitoring for men and women in groups and dyadic conflicts. *Journal of Applied Psychology* 91(2): 272–281.

Ford, C. 2007. *Women speaking up: Getting and using turns in workplace meetings*. New York: Palgrave/Macmillan. doi:10.1037/0021-9010.91.2.272

Foschi, M. 1996. Double standards in the evaluation of men and women. *Social Psychology Quarterly* 59: 237–254

Fox, Mary Frank. 2003. Gender, faculty, and doctoral education in science and engineering. In Lilli S. Hornig, ed. *Equal rites, unequal outcomes: Women in American research universities*. New York: Kluwer Academic. doi:10.2307/2787021

Fox, Mary Frank and Carol Colatrella. 2006. Participation, performance, and advancement of women in academic science and engineering: what is at issue and why. *Journal of Technology Transfer* 31(3): 377–386.

Freeman, S.T. 2002. *Conversations with powerful African-American women leaders: Inspiration, motivation, and strategy*. Washington, D.C.: AASBEA. doi:10.1007/s10961-006-7209-x

Frehill, Lisa M. et al., 2006. *Using program evaluation to ensure the success of your advance program advance: Institutional transformation working group*. http://www.advance.vt.edu/Measuring_Progress/Toolkits/Advance_Program_Evaluation_Toolkit_08May06.pdf

Gibson, Cristina B. 1995. An investigation of gender differences in leadership across four countries. *Journal of International Business Studies* 26.

Glazer-Raymo, Judith. 1999. *Shattering the myths: Women in academe*. Baltimore: The Johns Hopkins University Press. doi:10.1057/palgrave.jibs.8490847

Goleman, D. 1998. What makes a leader? *Harvard Business Review*. November–December.

Goleman, D. 1998. *Working with emotional intelligence*. New York: Bantam.

Gosling, J. and H. Mintzberg. 2003. The five minds of a manager. *Harvard Business Review* 81.11: 54–63.

Government Accountability Office. 2004. *Gender issues: Women's participation in the sciences has increased, but agencies need to do more to ensure compliance with Title IX* (GAO-04-639). Washington, DC: U.S. Government Accountability Office.

Gunter, R. and Stambach, A. 2005. Differences in men and women scientists' perceptions of workplace climate. *Journal of Women and Minorities in Science and Engineering* 11(1) 97–116.

Gunter, R. and Stambach, A. 2003. As balancing act and as game: How women and men science faculty experience the promotion process. *Gender Issues*, 21(1), pp. 24–42. doi:10.1615/JWomenMinorScienEng.v11.i1.60

Gurin, P., E.L. Dey, S. Hurtado, G. Gurin. 2002. Diversity and higher education: Theory and impact on educational outcomes. *Harvard Educational Review* 71, 3: 332–366. doi:10.1007/s12147-003-0020-1

Gurin, Patricia, Biren Nagda and Gretchen Lopez. 2004. The benefits of diversity in education for democratic citizenship. *Journal of Social Issues* 60.1:17–34.

Gustafson, J.E. 2000. *Some leaders are born women! Stories and strategies for building the leader within you.* Detroit, MI: Leaders Publications.

Halpern, D., C. Benbow, D. Geary, R. Gur, J. Shibley Hyde, and M.A. Gernsbacher. 2007. The science of sex differences in science and mathematics. *Psychological Science in the Public Interest* 8.1:1–51.

Handelsman, J., N. Cantor, M. Carnes, N. Hopkins, C. Marrett, D. Denton, E. Fine, S. Rosser, J. Sheridan, and V. Valian. 2005. More women in science. *Science* 309(5738):1190–1191. doi:10.1111/j.1529-1006.2007.00032.x

Hannum, Kelly M., Jennifer W. Martineau, and Claire Reinelt, eds. 2007. *The handbook of leadership development evaluation.* Center for Creative Leadership. doi:10.1126/science.1113252

Hanson, S.L. 2004. African American women in science: Experiences from high school through the post-secondary years and beyond. *NWSA Journal* 16.1:96.

Harding, S. 1986. The science question in feminism. Ithaca, NY: Cornell University Press. doi:10.1353/nwsa.2004.0033

Harding, S. 1998. *Is science multi-cultural? Postcolonialisms, feminisms, and epistemologies.* Bloomington: Indiana University Press.

Harrison, P. 2000. *A seat at the table: An insider's guide for America's new women leaders.* San Diego, CA: Master Media.

Harvard University. 2005. Report of the Task Force on Women Faculty, http://www.news.harvard.edu/gazette/daily/2005/05/women-faculty.pdf.

Haulbert, K. M. 1997. *Women as leaders.* New York: Marc Publishers.

Heilman, Madeline and T.G. Okimoto. 2007. Why are women penalized for success at male tasks?: The implied communality deficit. *Journal of Applied Psychology* 92(1): 81–92.

Heilman, M.E., A.S. Wallen, D. Fuchs, and M.M. Tamkins. 2004. Penalties for success: Reactions to women who succeed at male gender-typed tasks. *Journal of Applied Psychology* 89(3): 416–427. doi:10.1037/0021-9010.92.1.81

Heilman, M.E. 2001. Description and prescription: How gender stereotypes prevent women's ascent up the organizational ladder. *Journal of Social Issues* 57: 657–674 doi:10.1037/0021-9010.89.3.416

Hefferman, M. 2002. The female CEO ca. 2002. *Fast Company, 61:* 9, 58–60, 62, 64, 66. August. doi:10.1111/0022-4537.00234

Helgesen, S. 1995. *Female advantage: Women's ways of leadership.* New York: Doubleday Currency.

Hennessey J., S. Hockfield, and S. Tilghman. 2005. Women and science: The real issue. *Boston Globe,* Feb. 12, http://www.boston.com/news/education/higher/articles/2005/02/12/women_and_science_the_real_issue/.

Hersey, P. 1984. *The situational leader.* New York: Warner Books.

Hersey, P. K. Blanchard, and D. Johnson, 2008. *Management of organizational behavior: Leading human resources.* Upper Saddle River, NJ: Pearson Education.

Hojat M., J.S. Gonnella and A.S. Caelleigh. 2003. Impartial judgment by the 'gatekeepers' of science: Fallibility and accountability in the peer review process. *Advances in Health Sciences Education* 8.1:75–96.

Hoppe, M. 1998. Cross-cultural issues in leadership development. In Cynthia McCauley, Russ Moxley and Ellen Van Elsor, eds. *The Center for Creative Leadership handbook of leadership development.* San Francisco, CA: Jossey-Bass.

Hornig, Lilli S., ed. 2003. *Equal rites, unequal outcomes: Women in American research universities.* Kluwer Academic Press.

Hoyt, Crystal L. 2005. The role of leadership efficacy and stereotype activation in women's identification with leadership. *Journal of Leadership & Organizational Studies* 11.4: 2–14.

Huy, Q.N. 2001. In praise of middle managers. *Harvard Business Review* 79.8: 73–79. doi:10.1177/107179190501100401

Hunt, J.G., and A. Ropo. 1995. Multi-level leadership: grounded theory and mainstream theory applied to the case of General Motors. *Leadership Quarterly* 6(3): 379–412.

Hurtado, S. 2007. Linking diversity with the educational and civic missions of higher education. *Review of Higher Education* 30.2 (Winter):185–196 doi:10.1016/1048-9843(95)90015-2

Hyde, J.S. 2005. The gender similarities hypothesis. *American Psychologist* 60: 581–592. doi:10.1353/rhe.2006.0070

Ibarra, Herminia. 1993. Personal networks of women and minorities in management: A conceptual framework. *The Academy of Management Review* 18.1 (Jan) 56–87: 56. doi:10.1037/0003-066X.60.6.581

Inglehart, R. and P. Norris. 2003. *Rising tide: Gender equality and cultural change.* New York: Cambridge University Press. doi:10.2307/258823

Itzin, C. 1995. *The gender culture in organizations.* In C. Itzin and J. Newman, eds. *Gender, culture and organizational change: Putting theory into practice.* London: Routledge.

Jackson, D., E. Engstrom and T. Emmers-Sommer. 2007. Think leader, think male and female: Sex vs. seating arrangement as leadership cues. *Sex Roles* 57.9–10: 713–723.

Jackson, Judy. 2004. The story is not in the numbers: academic socialization and diversifying the faculty. *NWSA Journal* 16(1): 172–185. doi:10.1007/s11199-007-9289-y

Jacobs, Jerry A. and Kathleen Gerson. 2004. *The time divide: Work, family, and gender inequality.* Cambridge, MA: Harvard University Press. doi:10.2979/NWS.2004.16.1.172

Jamieson, K. H. 1997. *Beyond the double bind: Women and leadership.* Cambridge, MA: Oxford University Press.

Jamison, L., and M. Dietz, eds. 1997. *Leadership: Is gender an issue: Women's leadership in the 21st century*. The Center for Strategic and International Studies, Washington, D.C.

Judy, R.W., D'amico, C. and Geipel, G.L. 1987. *Workforce 2020: Work and workers in the 21st century*. Indianapolis: Hudson Institute.

Kang, Kelly. 2003. *Characteristics of doctoral scientists and engineers in the United States: 2001*. NSF03-310. Arlington, VA: National Science Foundation, Division of Science Resources Statistics.

Kanter, R.M. 1977. *Men and women of the corporation*, New York: Basic Books.

Kellerman, Barbara. 2004. *Bad leadership: What it is, how it happens, why it matters*. Cambridge, MA: Harvard Business School Press.

Kelly, R.M., M.M. Hale, and J. Burgess. 1991. Gender and managerial/leadership styles: A comparison of Arizona public administrators. *Women and Politics*, 11: 19–39.

Kezar, Adrianna. 2000. Pluralistic leadership: incorporating diverse voices. *The Journal of Higher Education* (6) (Nov–Dec): 722–743. doi:10.1300/J014v11n02_02

King, Jaqueline and Gigi Gomez. 2007. *The American college president: 2007 Edition*, ACE Center for Policy Analysis. doi:10.2307/2649160

Klenke, K. 1996. *Women and leadership: A contextual perspective*. New York: Springer Publishing.

Knight, P.T. and P.R. Trowler. 2000. Department-level cultures and the improvement of learning and teaching. *Studies in Higher Education*. 25(1):69–83.

Knight, W.H. and M.C. Holen. 1985. Leadership and the perceived effectiveness of department chairpersons. *Journal of Higher Education*, 56.6: 677–690.

Koch, Sabine C., Rebecca Luft and Lenelis Kruse. 2005. Women and leadership—20 years later: A semantic connotation study. *Social Science Information* 44.1: 9–39. doi:10.2307/1981074

Kram, K. E. 1988. *Mentoring at work: Developmental relationships in organizational life*. New York: University Press of America. doi:10.1177/0539018405050433

Kruse, L. and M. Wintermantel. 1986. Leadership Ms.-qualified: The gender bias in everyday and scientific thinking. In C.F. Graumann and S. Moscovici, eds. *Changing conceptions of leadership*. New York: Springer-Verlag: 171–197.

Kotter, J. P. 1995. *Leading change*. Boston: Harvard Business School Press.

Kotter, J. P. 1996. Leading change: why transformation efforts fail. *Harvard Business Review*, 73(2), 59–67.

Kotter, J.P. 1995. *The new rules: How to succeed in today's post-corporate world*. New York: Free Press.

Kouzes, J. and B. Posner. 1987. *The leadership challenge: How to get extraordinary things done in organizations*. San Francisco, CA: Jossey-Bass.

Kouzes, J. and B. Posner. 1995. *The leadership challenge: How to keep getting extraordinary things done in organizations*. San Francisco, CA: Jossey-Bass.

Kraatz, M.S. 1998. Learning by Association? Inter-organizational Networks and Adaptation to Environmental Change. *Academy of Management Journal*, 41(6): 621–643.

Kristof, N. When women rule. 2008. *New York Times* (8 Feb). doi:10.2307/256961

Kuh, Charlotte. 2003. You've come a long way: Data on women doctoral scientists and engineers in research universities. In L. S. Hornig, ed. *Equal rites, unequal outcomes: Women in American research universities*. New York: Kluwer Academic.

Kuhn, Thomas. 1962. *The structure of scientific revolutions*. Chicago: University of Chicago Press.

Kunda, Z. and S.J. Spencer. 2003. When do stereotypes come to mind and when do they color judgment? A goal-based theory of stereotype activation and application. *Psychological Bulletin* 129: 522–544.

Lach, J. 1999. Minority women hit concrete ceiling. *American Demographics* 21(9):18–19. doi:10.1037/0033-2909.129.4.522

Layne, Peggy. 2004. Women in Engineering Leadership Summit. *Society of Women Engineers*. Fall: 34.

Library of Congress, THOMAS, S-568; H.R.5305: Women in Science and Technology Equal Opportunity Act, http://thomas.loc.gov/cgi-bin/bdquery/z?d096:SN00568:@@@L.

Lighty, J.S., J. Vance, and D. Niemeier. 2005. Women in Engineering Leadership Institute (WELI). *Proceedings of the 2005 WEPAN/NAMEPA Joint Conference*.

Lincoln, J. R., and J. Miller. 1979. Work and friendship ties in organizations: A comparative analysis of relational networks. *Administrative Science Quarterly*. 24: 181–199.

Lipsey, Richard, Peter Steiner, Douglas Purvis, Paul Courant. 1990. *Economics*. New York: Harper & Row. doi:10.2307/2392493

Long, J. Scott, ed., *From scarcity to visibility: Gender differences in the careers of doctoral scientists and engineers* (Washington, D.C.: National Academy Press, 2001).

Long, J.S. and M.F. Fox. 1995. Scientific careers: Universalism and particularism. *Annual Review of Sociology* 21: 45–71.

Luke, Nancy and Munshi Kaivan. 2004. Women as agents of change: Female incomes and household decisions in South India. Center for International Development, Harvard University, BREAD Working Paper No. 087. www.cid.harvard.edu/bread/papers/041604_Conference/0504conf/bread_munnar3.pdf. doi:10.1146/annurev.so.21.080195.000401

Madden, Margaret E. 2005. 2004 Division 35 Presidential Address: Gender and Leadership in Higher Education. *Psychology of Women Quarterly* 29.1: 3–14.

Madsen, Susan R. 2007. Women university presidents: Career paths and educational backgrounds. *Academic Leadership: The Online Journal* (5) 1. http://www.academicleadership.org/emprical_research/Women_University_Presidents_Career_Paths_and_Educational_Backgrounds.shtml doi:10.1111/j.1471-6402.2005.00162.x

Marra, R.M. and B. Bogue. 2004. AWE: A model for sustainable and profitable collaboration between disciplines. *Journal of Women and Minorities in Science and Engineering*, 10(3): 283–295.

Mason, Mary Ann, Angelica Stacy, Marc Goulden, Carol Hoffman, and Karie Frasch. 2005. University of California faculty family friendly edge: An initiative for tenure-track faculty at the University of California. http://ucfamilyedge.berkeley.edu/ucfamilyedge.pdf.

Massachusetts Institute of Technology. 1999. A study on the status of women faculty in science at MIT. *MIT Faculty Newsletter* 11(4), *http://web.mit.edu/fnl/women/women.html*.

McGuire, Gail M. 2002. Gender, race, and the shadow structure: A study of informal networks and inequality in a work organization. *Gender and Society*, 16.3 (Jun): 303–322.

McLeod, P.L., S.A. Lobel, and T.H. Cox. 1996. Ethnic diversity and creativity in small groups. *Small Group Research* 27: 248–265. doi:10.1177/0891243202016003003

Meyerson, D., and J.K. Fletcher. 2000. A modest manifesto for shattering the glass ceiling. *Harvard Business Review*, 78(1), 126–137. doi:10.1177/1046496496272003

Meier, Stephan. 2005. Conditions under which women behave less/more pro-socially than men: Evidence from two field experiments. John F. Kennedy School of Government Women and Public Policy Program (WAPPP) Working Paper Series. http://www.hks.harvard.edu/wappp/research/working/index.html

Nemeth, Charlan J. 1995. Dissent as driving cognition, attitudes, and judgments. *Social Cognition* 13: 273–291.

Miller, D.T., B. Taylor and M.L. Buck. 1991. Gender gaps: Who needs to be explained? *Journal of Personality and Social Psychology*, 61, 5–12.

Milem, J. F. 2003. The educational benefits of diversity: Evidence from multiple sectors. In Mitchell Chang et al., eds. *Compelling interest: Examining the evidence on racial dynamics in higher education*. Stanford: Stanford Education. doi:10.1037/0022-3514.61.1.5

Milem, J.F., M.J. Chang and A.L. Antonio. 2005. *Making diversity work on campus: A research-based perspective*. Washington, DC: Association of American Colleges and Universities.

Morrison, A.M. and M.A. Von Glinow. 1990. Women and minorities in management. *American Psychologist* 45: 200–208.

Morrison, A.M., R.P. White and E. Van Velsor. 1992. *Breaking the glass ceiling: Can women reach the top of America's largest corporations?* Reading, MA: Addison-Wesley. doi:10.1037/0003-066X.45.2.200

Murrell, K. 1997. Emergent theories of leadership for the next century: Towards relational concepts. *Organization Development Journal* 15.3: 35–42.

Naff, K.C. and J.E. Kellough. 2003. Ensuring employment equity: Are federal diversity programs making a difference? *International Journal of Public Administration* 26.12:1307–1336.

Nanus, Burt and Stephen M. Dobbs. 1999. *Leaders who make a difference: Essential strategies for meeting the nonprofit challenge.* San Francisco: Jossey-Bass. doi:10.1081/PAD-120024399

National Academies of Science (NAS/NAE/IOM). 2007. *Rising above the gathering storm: Energizing and employing America for a brighter economic future.* Washington, DC: The National Academies Press.

National Research Council. 2001. *From scarcity to visibility: Gender differences in the careers of doctoral scientists and engineers.* Washington, DC: National Academy Press.

National Research Council, Committee on Women in Science and Engineering. 2006. *To Recruit and advance: Women students and faculty in science and engineering*, National Academies Press, Washington, D.C.

National Research Council, Committee on Women in Science and Engineering. 2007. *Beyond bias and barriers: Fulfilling the potential of women in academic science and engineering.* National Academies Press, Washington, D.C.

National Science Board. 2006. *Science and engineering indicators*, 2006 (NSB 06-02). Arlington, VA: NSF, 2-5, http://www.nsf.gov/statistics/seind06/c2/c2h.htm.

National Science Foundation, Division of Science Resources Statistics, 2004. *Gender differences in the careers of academic scientists and engineers*, NSF 04-323, Project Officer, Alan I. Rapoport (Arlington, VA).

National Science Foundation, Division of Science Resources Statistics, 2007. *Women, minorities, and persons with disabilities in science and engineering: 2007*, NSF 07-315 (Arlington, VA; February 2007). http://www.nsf.gov/statistics/wmpd.

Nelson, D.J. 2005. *A national analysis of diversity in science and engineering faculties at research universities.* Norman, OK (revised Oct 2007). Available online: http://cheminfo.chem.ou.edu/~djn/diversity/briefings/Diversity%20Report%20Final.pdf.

Niemeier, D.A. 2005a, *Conference Evaluation. Summary of Findings: Pre-Conference and Post-Conference Surveys.* NSF Women in Engineering Leadership Development Conference, Cocoa Beach, FL, Oct 2005.

Niemeier, D.A. 2005b. *Conference Evaluation. Summary of Findings: Pre-Conference and Post-Conference Surveys.* NSF Women in Engineering Leadership Institute: The Advanced Leadership Workshop, Syracuse, NY, Apr 2005.

Niemeier, D.A., 2004a. *Conference Evaluation, Summary of Findings: Pre-Conference and Post-Conference Surveys.* NSF Women in Engineering Leadership Development Conference, Salt Lake City, Mar 2004.

Niemeier, D.A. 2004b. *Conference Evaluation, Summary of Findings: Pre-Conference and Post-Conference Surveys.* NSF Women in Engineering Leadership Summit, Storrs, CT, Jun 2004. www.weli.eng.iastate.edu/Resources/weli_reports.html.

Niemeier, D.A. and Gonzalez, C. 2004. Breaking into the guildmaster's club: What we know about women science and engineering department chairs at AAU universities. *NWSA Journal*, 16 (1), 157–171.

Nichols, Nancy. 1993. Whatever happened to Rosie the Riveter? *Harvard Business Review* 71.4 (Jul–Aug) 54–57.

Nyquist, L.V. and J.T. Spence. 1986. Effects of dispositional dominance and sex role expectations on leadership behaviors. *Journal of Personality and Social Psychology* 50: 87–93.

Oakley, J.G. 2000. Gender-based barriers to senior management positions: Understanding the scarcity of female CEOs. *Journal of Business Ethics*, 27(4), 321–334. doi:10.1037/0022-3514.50.1.87

Olson, M. 1965. *The logic of collective action: Public goods and the theory of groups.* Cambridge, MA: Harvard University Press.

Olsson, S. 2000. Acknowledging the female archetype: Women managers' narratives of gender. *Women in Management Review* 15.5/6: 296–302.

Ottinger, Cecilia and Robin Sikula. 1993. Women in higher education: Where do we stand? *Research Briefs* 4.2. American Council on Education, Washington, D.C. doi:10.1108/096494 20010372959

Oyster, C.K. 1992. Perceptions of power. *Psychology of Women Quarterly* 16: 527–533.

Parker, Patricia, S. 2006. *Race, gender, and leadership: Re-envisioning organizational leadership from the perspectives of African American women executives.* London: Lawrence Erlbaum Associates. doi:10.1111/j.1471-6402.1992.tb00273.x

Perusek, Anne. 2004. Fostering leaders. *Society of Women Engineers*. Summer.

Perusek, A.M. 2008. The leaky science and engineering pipeline: how can we retain more women in academia and industry? *SWE - Magazine of the Society of Women Engineers*. Winter: 20–22.

Pfeffer, Jeffrey. 1977. The ambiguity of leadership. *Academy of Management Review* 2.

Phelps, E. 1972. The statistical theory of racism and sexism. *American Economic Review* 62: 659–661. doi:10.2307/257611

Podolny, J.M. and J. N. Baron, 1997. Resources and relationships: Social networks and mobility in the workplace. *American Sociological Review* 62: 673–693.

Posner, B. and Brodsky, B. 1992. A leadership development instrument for college students. *Journal of College Student Development* 33.3: 231–237. doi:10.2307/2657354

Powell, G.N., D.A. Butterfield, and J.D. Parent. 2002. Gender and managerial stereotypes: Have the times changed? *Journal of Management* 28: 177–193: 188. doi:10.1016/S0149-2063(01)00136-2

Riley, Donna and Caroline Baillie. 2008. *Engineering and social justice*. Morgan & Claypool Publishers.

Rosenbach, W., Sashkin, M., and Harburg, F. 1996. *The leadership profile: On becoming a better leader.* Seabrook, MD: Ducochon Press.

Rosener, Judith. 1990. Ways women lead. *Harvard Business Review*, 68, 119–125.

Rosener, J.B. 1995. *America's competitive secret: Utilizing women as management strategy.* New York: Oxford University Press.

Rosener, J.B. 1995. *America's competitive secret: Utilizing women as management strategy.* New York: Oxford University Press.

Rosser, S.V. and E. O'Neil Lane. 2002. Key barriers for academic institutions seeking to retain female scientists and engineers: Family-unfriendly policies, low numbers, stereotypes, and harassment. *Journal of Women and Minorities in Science and Engineering* 8: 161–189.

Rosser, Sue V. 2004. *The science glass ceiling: Academic women scientists and the struggle to succeed.* New York: Routledge.

Rosser, Sue V. 2006. Senior women scientists: Overlooked and understudied? *Journal of Women and Minorities in Science and Engineering* 12(4): 275–293. doi:10.1615/JWomenMinorScienEng.v12.i4.20

Rosser, Sue V. 2005. Women and technology through the lens of feminist theories. *Frontiers: A Journal of Women's Studies*, 26.1: 1–23.

Rosser, S. 2002. Twenty-five years of NWSA: Have we built the two-way streets between Women's Studies and Women in Science and Technology? *NWSA Journal* (14.1).

Rosser, V.J., L.K. Johnsrud and R.H. Heck. 2003. Academic deans and directors: Assessing their effectiveness from individual and institutional Perspectives. *The Journal of Higher Education* 74.1 (January/February) 1–25. doi:10.1353/jhe.2003.0007

Rossiter, M.W. 1995. *Women scientists in America: Before affirmative action* 1940–1972. Baltimore, MD: Johns Hopkins Press.

Rost, J.C. 1993. Leadership development in the new millennium. *The Journal of Leadership Studies*: 91–110.

Rost, J.C. 1991. *Leadership for the twenty-first century.* Westport, CT: Praeger.

Ruderman, Marian N. and Patricia J. Ohlott. 2002. *Standing at the crossroads: Next steps for high-achieving women.* Center for Creative Leadership.

Rudman, L. and K. Fairchild. 2004. Reactions to counterstereotypic behavior: The role of backlash in cultural stereotype maintenance. *Journal of Personality and Social Psychology*, 87.2: 157–176. doi:10.1037/0022-3514.87.2.157

Rudman, L. and S. Kilianski. 2000. Implicit and explicit attitudes toward female authority. *Personality and Social Psychology Bulletin* 26.11:1315–1328. doi:10.1177/0146167200263001

Safferstone, M.J. 2005. Organizational leadership: Classic works and contemporary perspectives. *CHOICE: Current Reviews for Academic Libraries*, 42(6), 959–975.

Schein, V.E. 2001. A global look at psychological barriers to women's progress in management. *Journal of Social Issues* 57: 675–688. doi:10.1111/0022-4537.00235

Schein, V.E. 1973. The relationship between sex role stereotypes and requisite management characteristics. *Journal of Applied Psychology* 57: 95–100.

Settles, Isis H, Cortina, L.M. Malley, J. and Stewart A.J. 2006. The climate for women in academic science: The good, the bad, and the changeable. *Psychology of Women Quarterly* 30(1): 47–58. doi:10.1111/j.1471-6402.2006.00261.x

Shackelford, S., W. Wood, and S. Worchel. 1996. Behavioral styles and the influence of women in mixed-sex groups. *Social Psychology Quarterly* 59: 284–293. doi:10.2307/2787024

Sharpe, R. 2000. As leaders, women rule: New studies find that female managers outshine their male counterparts in almost every measure. *Business Week* (Nov 20). http://www.businessweek.com/common_frames/ca.htm?/2000/00_47/b3708145.htm.

Sheridan, J., Handelsman, J., and Carnes, M. 2004. Assessing 'readiness to embrace diversity': an application of the trans-theoretical model of behavioral change. PowerPoint presentation at the American Sociological Association meeting, session entitled Workplace Diversity, San Francisco, CA.

Shrier, Diane K., Alyssa N. Zucker, Andrea E. Mercurio, Laura J. Landry, Michael Rich, Lydia A. Shrier. 2007. Generation to generation: Discrimination and harassment experiences of physician mothers and their physician daughters. *Journal of Women's Health*, 16.6 (July 1): 883–894. doi:10.1089/jwh.2006.0127

Smith, Walter S. and Thomas Owen Erb. 1986. Effect of women science career role models on early adolescents' attitudes toward scientists and women in science. *Journal of Research in Science Teaching* 23.8 (Nov): 667–676. doi:10.1002/tea.3660230802

Smith, Daryl et al. 1997. *Diversity works: The emerging picture of how students benefit.* Association of American Colleges and Universities, Washington D.C.

Smith-Doerr, Laurel, 2004. Flexibility and fairness: Effects of the network form of organization on gender equity in life science careers. *Sociological Perspectives* 47.1: 25–54. doi:10.1525/sop.2004.47.1.25

Sonnert G. and G. Holton. 1996. The career patterns of men and women scientists. *American Scientist* (Jan).

South, S. J., C.M. Bonjean, W.T. Markham and J. Corder. 1982. Social structure and intergroup interaction: Men and women of the federal bureaucracy. *American Sociological Review.* 47: 587–599. doi:10.2307/2095160

Spelke, Elizabeth. 2005. Sex differences in intrinsic aptitude for mathematics and science?: A critical review. *American Psychologist*, 60.9 (Dec): 950–958. doi:10.1037/0003-066X.60.9.950

Spencer, S.J., C.M. Steele, D.M. Quinn. 1999. Stereotype threat and women's math performance. *Journal of Experimental Social Psychology* 35: 4–28. doi:10.1006/jesp.1998.1373

Stead, Valerie. 2005. Mentoring: A model for leadership development? *International Journal of Training and Development* 9.3:170–184. doi:10.1111/j.1468-2419.2005.00232.x

Steele, C.M. 1997. A threat in the air: How stereotypes shape intellectual identity and performance. *American Psychologist* 52:613–629. doi:10.1037/0003-066X.52.6.613

Steinpreis, Rhea; Katie A. Ander and Dawn Ritzke. 1999. The impact of gender on the review of the curricula vitae of job applicants and tenure candidates: A national empirical study. *Sex Roles* 41: 509–528.

Stogdill, R.M. 1974. *Handbook of leadership: A survey of theory and research*. New York: Free Press.

Strom, S. 2008. Gift to teach business to Third-World women. *New York Times*, March 6.

Swamy, Anand, Stephen Knack, Young Lee, and Omar Azfar. 2001. Gender and corruption. *Journal of Development Economics* 64 (Feb) 25–55. doi:10.1016/S0304-3878(00)00123-1

Swoboda, M. and J. Vanderbosch. 1986. The society of outsiders: Women in administration. In P. Farrant, ed. *Strategies and attitudes. Women in educational administration*. National Association for Woman Deans, Administrators and Counselors, Washington, D.C.

Syracuse University *Senate Committee on Diversity*. 2007. *Senate Committee on Diversity Recommendations in Faculty Retention* (Feb), http://universitysenate.syr.edu/diversity/diversity-21mar07.pdf.

Sczesny, S. 2003. The perception of leadership competence by female and male leaders. *Zeitscrift fur Socialpsychologie* 34:133–145.

Tichy, N.M. 1981. Networks in organizations. In P.C. Nystrom and W.H. Starbuck, eds. *Handbook of organization design*. New York: Oxford University Press: 225–248.

Thomas, D. A. 1990. The impact of race on managers' experience of developmental relationships (mentoring and sponsorship): An intra-organizational study. *Journal of Organizational Behavior*. 2: 479–492. doi:10.1002/job.4030110608

Thompson, M.D. 2000. Gender, leadership orientation and effectiveness: Testing the theoretical model of Bolman & Deal and Quinn. *Sex Roles* 42.11/12: 969–992.

Trix, Frances and Psenka, Carolyn. 2003. Exploring the color of glass: Letters of recommendation for female and male medical faculty. *Discourse & Society* 14(2): 191–220. doi:10.1177/0957926503014002277

Tsui, A.S. and C.A. O'Reilly. 1989. Beyond simple demographic effects: The importance of relational demography in superior-subordinate dyads. *Academy of Management Journal*. 32: 402–423. doi:10.2307/256368

United Nations Proposal Concerning a Definition of the Term 'Minority.' 1985. UN Document E/CN.4/Sub.2/1985/31.

United Nations. 2005. *The World's Women 2005: Progress in Statistics* (ST/ESA/STAT/SER.K/17) Demographic and Social Statistics Branch. New York, NY. http://unstats.un.org/unsd/demographic/products/indwm/ww2005/tab6.htm.

U.S. Glass Ceiling Commission. 1995. *A Solid Investment: Making Full Use of the Nation's Human Capital*. Final Report of the Commission, Washington, D.C., U.S. Government Printing Office. http://digitalcommons.ilr.cornell.edu/key_workplace/120;

Valian, V. 1999. *Why so slow: The advancement of women*. Cambridge, MA: The MIT Press.

van Engen, M.L., R. van der Leeden and T. Willemsen. 2001. Gender, context and leadership styles: A field study. *Journal of Occupational and Organizational Psychology* 74: 581–598. doi:10.1348/096317901167532

Vecchio, R.P. 2002. Leadership and the gender advantage. *Leadership Quarterly* 13: 643–672. doi:10.1016/S1048-9843(02)00156-X

Walker, D.E. 1979. *The effective administrator: A practical approach to problem solving, decision making, and campus leadership*. San Francisco: Jossey-Bass.

Wanda, E. Ward. 2001. The success of female scientists in the 21st century: An American perspective. In *Women in scientific careers: Unleashing the potential*: 149–154. OECD Publishing.

Weisgram, Erica S. and Bigler, Rebecca S. 2007. Effects of learning about gender discrimination on adolescent girls' attitudes toward and interest in science. *Psychology of Women Quarterly* 31(3): 262–269. doi:10.1111/j.1471-6402.2007.00369.x

Wenniger, Mary D. and Mary H. Conroy, eds. 2001. *Gender equity or bust: On the road to campus leadership with women in higher education*. San Francisco: Jossey-Bass.

West, C., and Zimmerman, D.H. 1991. Doing gender. In S. Farrell, ed. *The social construction of gender*. Newbury Park, CA: Sage.

Womack, R.B. 1996. Measuring the leadership styles and scholarly productivity of nursing department chairpersons. *Journal of Professional Nursing* 12.3:133–140. doi:10.1016/S8755-7223(96)80036-6

Wood, W. and S.J. Karten. 1996. Sex differences in interaction style as a product of perceived sex differences in competence. *Journal of Personality and Social Psychology* 50: 341–347. doi:10.1037/0022-3514.50.2.341

Xie, Y. and K.A. Shauman. 2003. *Women in science: Career processes and outcomes*. Cambridge, MA: Harvard University Press.

Yukl, G. 2002. *Leadership in organizations, 5th* edn. Upper Saddle River, N.J.: Prentice-Hall.

Yukl, G. 1999. An evaluation of conceptual weaknesses in transformational and charismatic leadership theories. *The Leadership Quarterly* 10.2 (Summer): 285–305. doi:10.1016/S1048-9843(99)00013-2

Yukl, G., P.J. Guinan, and D. Sottolano. 1995. Influence tactics used for different objectives with subordinates, peers, and superiors. *Group and Organization Management* 20: 272– 296.

Yukl, G. and C.E. Seifert. 2002. Preliminary validation research on the extended version of the influence behavior questionnaire. Paper presented at the Society for Industrial and Organizational Psychology Annual Conference.

Zane, N. 1999. Gender and leadership: The need for 'public talk' in building an organizational change agenda. *Diversity Factor* 7.3: 16–21.

Zuckerman, H., J.R. Cole and J.T. Bruer, eds. 1981. *The outer circle: Women in the scientific community*. New York: Norton.

Author Biography

Corri Zoli is a postdoctorate research fellow and program manager for the Women in Science and Engineering (WISE) Program at the L.C. Smith College of Engineering and Computer Science at Syracuse University in Syracuse, New York. Her Ph.D. in cultural studies is from the Department of English at Syracuse University, she has a Certificate of Advanced Study (CAS) in Women's Studies, and she is in the process of completing a policy-focused M.A. in international relations from the Maxwell School of Syracuse University. Her research has focused on women and gender in institutions, the history of science and technology in relation to society, diversity and globalization, and the role of culture and identity in international relations, including security studies. Her policy interests include higher education, science and technology law and policy, the social and cultural impact of identity and diversity, and gender and conflict in the international context. She has several publications in these areas and has received awards for her writing and teaching.

Shobha K. Bhatia is Laura J. and L. Douglas Meredith professor and professor of civil and environmental engineering at Syracuse University in Syracuse, New York. She has made significant contributions to both engineering research and engineering education. Her engineering research has focused on the application of geosynthetics and natural materials in waste containment, road and building construction, and erosion control. She has over 80 publications; has received funding from the National Science Foundation (NSF), the United States Environmental Protection Agency (USEPA), the New York State Department of Transportation (NYSDOT), and many other private organizations; has participated in national and international conferences; and has served in numerous capacities, such as vice president of the North American Geosynthetics Society (NAGS), member of the Technical Coordination Council (TCC), and member of the International Activity Council of the Geo-Institute of the American Society of Civil Engineers (ASCE). She has also been extensively involved in engineering education. She is co-director of the Women in Science and Engineering (WISE) initiative at Syracuse University. She has also been part of national initiatives to increase the number of women in leadership positions in academia through her projects funded by the NSF ADVANCE program. She played an important role in the NSF-funded Engineering Education Scholar Program, which was designed to prepare young faculty for academic careers. She is also the recipient of a NSF Faculty Achievement Award for Women for excellence in research and leadership in training future engineers and has received several national and international awards, including the International Network for Engineering Education and Research (iNEER) Award for Excellence in Fostering Sustained and Unique Collaborations in International Research and Education.

Valerie Davidson is the National Science and Engineering Research Council (NSERC) and Hewlett-Packard Canada Chair for Women in Science and Engineering (Ontario) and professor of biological engineering at the School of Engineering at the University of Guelph, Ontario, Canada. She has made significant contributions to both engineering research—in the areas of food science—and engineering education. As NSERC/HP Chair, she has undertaken such initiatives as encouraging girls in elementary and secondary schools to study science and engineering (S&E) and consider careers in these areas, improving retention of women in undergraduate and graduate programs and early careers related to S&E, and developing a provincial network to help women make informed decisions to pursue S&E from early education to postsecondary programs to early careers, including academic careers. In her research, Valerie has established a strong interdisciplinary program in food and biological engineering, with an emphasis on the applications of fuzzy mathematics and statistical methods to process control and decision-support systems. From 1990 to 1992, Valerie served as a member of the Canadian Committee on Women in Engineering. In 2002, she was a corecipient of the Canadian Council of Professional Engineers (CCPE) Award for Support of Women in Engineering, an award that recognizes noteworthy support of women in the engineering profession and engineering excellence. In addition, she is an active member of the Women in Engineering Leadership Institute (WELI).

Kelly Rusch is the associate dean of Women and Minorities and the Formosa Plastics Associate Professor of civil and environmental engineering at Louisiana State University (LSU). From 2000 to 2006, she was also the director of the Institute for Ecological Infrastructure Engineering at LSU. Her research interests include continuous growth/process control technology development for aquaculture production systems, water quality management and modeling of surface waters and aquaculture systems, wastewater treatment using natural wetlands, and industrial waste stabilization. She has numerous projects and an extensive number of articles on these and related subjects, and she has been the recipient of multiple federally funded grants in these areas. Additionally, she has made significant contributions to engineering education and to advancing the representation and participation of women and minorities in science and engineering. She has received the LSU Distinguished Faculty Award for a sustained record of excellence in teaching, research, and service and the Louisiana Engineering Foundation (LEF) Faculty Professionalism Award in 2005. She is also the recipient of an Achievement Award from the Department of Civil and Environmental Engineering at LSU in 2002. She has been an active member of the Women in Engineering Leadership Institute (WELI).

Printed in the United States
by Baker & Taylor Publisher Services